工业和信息化普通高等教育"十二五"规划教材立项项目

21世纪高等学校计算机规划教材

21st Century University Planned Textbooks of Computer Science

C语言程序设计实验指导（第2版）

The Practice of C Programming (2nd Edition)

马晓艳 许婷婷 马华 主编
张兰华 李聪 孙静 副主编

高校系列

人民邮电出版社

北京

图书在版编目（CIP）数据

C语言程序设计实验指导 / 马晓艳，许婷婷，马华主编． -- 2版． -- 北京：人民邮电出版社，2015.9
21世纪高等学校计算机规划教材．高校系列
ISBN 978-7-115-39721-8

Ⅰ．①C… Ⅱ．①马… ②许… ③马… Ⅲ．①C语言－程序设计－高等学校－教学参考资料 Ⅳ．①TP312

中国版本图书馆CIP数据核字(2015)第165780号

内 容 提 要

本书是《C语言程序设计基础》（第2版）配套使用的实验指导教材。

全书分为4个部分：第1部分为Visual C++6.0上机指导，详尽介绍了VC++ 6.0的编程环境、C语言程序调试方法和编程风格。第2部分为基本实验，共安排了15个实验，在第1版的基础上扩展了实验题目，以进一步巩固各章节的知识点。第3部分为综合性实验，共安排了3个实验，通过综合实验让学生熟练掌握结构化程序设计的思想和方法，培养学生综合应用知识的能力。第4部分为提高性实验，共安排了5个实验，目的是希望读者巩固C语言的基本知识，加强多知识点交叉实验的训练。最后在附录中给出了教材课后习题参考答案及提高性实验的参考程序。

本书实验安排恰当，例题、习题丰富，分析透彻，与主教材配套，便于自学。本书可作为高等学校"计算机程序设计"课程的实验教材，也可作为各类考试和社会读者的自学辅助用书。

◆ 主　编　马晓艳　许婷婷　马　华
　　副主编　张兰华　李　聪　孙　静
　　责任编辑　许金霞
　　责任印制　沈　蓉　彭志环

◆ 人民邮电出版社出版发行　北京市丰台区成寿寺路11号
　　邮编　100164　电子邮件　315@ptpress.com.cn
　　网址　http://www.ptpress.com.cn
　　北京天宇星印刷厂印刷

◆ 开本：787×1092　1/16
　　印张：10.75　　　　　2015年9月第2版
　　字数：281千字　　　2024年8月北京第8次印刷

定价：28.00元

读者服务热线：(010)81055256　印装质量热线：(010)81055316
反盗版热线：(010)81055315

前 言

　　C 语言是目前世界上最流行、使用最广泛的高级程序设计语言之一，同时又是一门实践性很强的课程。为了提高本课程的教学质量，改善 C 语言难讲、难学、难以掌握的现状，我们配合教材《C 语言程序设计基础》(第 2 版)，编写了《C 语言程序设计实验指导》(第 2 版)。

　　我们在第 1 版的基础上增加了 Visual C++6.0 上机指导、C 语言程序的调试方法和 C 语言编程风格等内容，扩充了基本实验内容；还增加了提高性实验并附有参考程序。

　　本书共分为 4 部分。

　　第 1 部分为 Visual C++6.0 上机指导，首先介绍了 Visual C++6.0 开发环境，然后介绍了程序调试的方法，最后介绍了 C 语言的编程风格。

　　第 2 部分为基本实验。共安排了 15 个实验项目，每个实验项目包括：实验项目名称、实验目的和实验内容。实验目的指出对学生的知识要求和能力要求，详细指明了应该达到的具体目标。实验内容则包含 3 部分的内容：验证性实验、改错性实验和设计性实验。验证性实验训练学生阅读程序，熟悉相应章节的知识点，体会良好的程序书写风格。改错性实验主要训练学生根据出错信息运用程序调试方法进行程序调试，提高学生的程序调试能力。设计性实验给出题目和部分题目的解题提示，让学生自己设计算法，并用 N-S 图描述，然后编程并调试，提高学生运用知识解决问题的能力。

　　第 3 部分为综合性实验，通过综合性实验的设计，让学生熟练使用结构化程序设计思想和方法，培养学生综合应用知识的能力。

　　第 4 部分为提高性实验。希望读者通过练习，巩固 C 语言的基本知识，加强多知识点交叉实验的训练，可使初学者能够快速掌握 C 语言程序设计。

　　最后在附录中给出了主教材中每章课后习题的参考答案以及提高性实验的参考答案。

　　程序设计是创造性的劳动，它需要编程人员的全身心投入，需要充分发挥主观能动性。书中的程序都是编者精心选取和编写的，特别是部分改错题都给出了错误提示行，部分程序设计题给出了详尽的算法提示、编程思路，并且全部在 Visual C++ 6.0 环境下调试通过。

　　本书虽然是《C 语言程序设计基础》(第 2 版)教材的配套参考书，但也可以单独使用，有些程序还具有一定的实用性，相信读者会从中得到启发。

编　者

2015 年 4 月

目 录

第1部分 Visual C++6.0 上机指导 1

1.1 Visual C++6.0 开发环境 1
1.1.1 Visual C++6.0 集成开发环境简介 1
1.1.2 启动 Visual C++6.0 集成开发环境 1
1.1.3 开始一个新程序 1

1.2 C语言程序调试方法 6
1.2.1 错误的类型 6
1.2.2 查错的方法 7
1.2.3 Visual C++ 6.0 中常用的程序调试工具 7

1.3 C语言的编程风格 8
1.3.1 编程风格 8
1.3.2 变量名、函数名的命名规则 9
1.3.3 注释 10

第2部分 基本实验 11

一、实验要求 11
二、实验报告的内容 11

实验1 初识C程序 11
一、实验目的 11
二、实验内容 12
三、实验注意事项 14

实验2 C语言基础 14
一、实验目的 14
二、实验内容 14
三、实验注意事项 17

实验3 顺序结构程序设计 17
一、实验目的 17
二、实验内容 18
三、实验注意事项 22

实验4 选择结构程序设计 22
一、实验目的 22
二、实验内容 22
三、实验注意事项 30

实验5 循环结构程序设计(一) 31

一、实验目的 31
二、实验内容 31
三、实验注意事项 37

实验6 循环结构程序设计(二) 37
一、实验目的 37
二、实验内容 37
三、实验注意事项 45

实验7 函数（一） 45
一、实验目的 45
二、实验内容 45
三、实验注意事项 58

实验8 函数（二） 58
一、实验目的 58
二、实验内容 58
三、实验注意事项 63

实验9 数组（一） 64
一、实验目的 64
二、实验内容 64
三、实验注意事项 76

实验10 数组（二） 76
一、实验目的 76
二、实验内容 77
三、实验注意事项 85

实验11 指针（一） 85
一、实验目的 85
二、实验内容 85
三、实验注意事项 90

实验12 指针（二） 90
一、实验目的 90
二、实验内容 91
三、实验注意事项 97

实验13 结构体 98
一、实验目的 98
二、实验内容 98
三、实验注意事项 102

实验14 文件 102

一、实验目的 …………………………… 102
　　二、实验内容 …………………………… 102
　　三、实验注意事项 ……………………… 106
　实验15　链表 …………………………… 107
　　一、实验目的 …………………………… 107
　　二、实验内容 …………………………… 107
　　三、实验注意事项 ……………………… 107

第3部分　综合性实验 …………………… 108

　实验1　学生成绩管理 …………………… 108
　　一、实验要求 …………………………… 108
　　二、实验提示 …………………………… 108
　　三、参考程序 …………………………… 109
　实验2　约瑟夫环问题 …………………… 118
　　一、实验要求 …………………………… 118
　　二、实验提示 …………………………… 119
　　三、参考程序 …………………………… 119
　实验3　双向链表的综合运算 …………… 120
　　一、实验要求 …………………………… 120
　　二、实验提示 …………………………… 120
　　三、参考程序 …………………………… 120

第4部分　提高性实验 …………………… 124

　提高性实验1 ……………………………… 124
　　一、填空型实验 ………………………… 124
　　二、改错型实验 ………………………… 125
　　三、设计型实验 ………………………… 126
　提高性实验2 ……………………………… 127
　　一、填空型实验 ………………………… 127
　　二、改错型实验 ………………………… 128
　　三、设计型实验 ………………………… 130
　提高性实验3 ……………………………… 131

　　一、填空型实验 ………………………… 131
　　二、改错型实验 ………………………… 132
　　三、设计型实验 ………………………… 134
　提高性实验4 ……………………………… 135
　　一、填空型实验 ………………………… 135
　　二、改错型实验 ………………………… 136
　　三、设计型实验 ………………………… 138
　提高性实验5 ……………………………… 139
　　一、填空型实验 ………………………… 139
　　二、改错型实验 ………………………… 140
　　三、设计型实验 ………………………… 141

附录　参考答案 …………………………… 143

　配套教材课后习题参考答案 …………… 143
　　第1章　C语言概述 …………………… 143
　　第2章　基本C语言程序设计 ………… 143
　　第3章　选择结构程序设计 …………… 144
　　第4章　循环结构程序设计 …………… 148
　　第5章　函数 …………………………… 149
　　第6章　数组 …………………………… 150
　　第7章　指针 …………………………… 157
　　第8章　结构体 ………………………… 159
　　第9章　文件 …………………………… 162
　提高性实验参考答案 …………………… 162
　　提高性实验1参考答案 ………………… 162
　　提高性实验2参考答案 ………………… 163
　　提高性实验3参考答案 ………………… 164
　　提高性实验4参考答案 ………………… 164
　　提高性实验5参考答案 ………………… 165

参考文献 …………………………………… 166

第 1 部分
Visual C++6.0 上机指导

1.1 Visual C++6.0 开发环境

1.1.1 Visual C++6.0 集成开发环境简介

Visual C++ 6.0，简称 VC 或者 VC6.0，是微软推出的一款 C++编译器，将"高级语言"翻译为"机器语言（低级语言）"的程序。Visual C++是一个功能强大的可视化软件开发工具。自 1993 年 Microsoft 公司推出 Visual C++1.0 后，随着其新版本的不断问世，Visual C++已成为专业程序员进行软件开发的首选工具。虽然微软公司推出了 Visual C++.NET（Visual C++7.0），但它的应用有很大的局限性，只适用于 Windows 2000、Windows XP 和 Windows NT4.0，所以实际中，更多的是以 Visual C++6.0 为平台。Visual C++6.0 不仅是一个 C++ 编译器，而且是一个基于 Windows 操作系统的可视化集成开发环境（integrated development environment，IDE）。Visual C++6.0 由许多组件组成，包括编辑器、调试器以及程序向导 AppWizard、类向导 Class Wizard 等开发工具。这些组件通过一个名为 Developer Studio 的组件集成为和谐的开发环境。

1.1.2 启动 Visual C++6.0 集成开发环境

安装好微软的 Visual Studio 6.0 后，在系统的"开始"菜单的"程序"中可以启动 Visual C++6.0，见图 1-1，也可以在桌面上为菜单项 Visual C++6.0 建立一个快捷图标，如图 1-2.所示。可选择这两个方法之一启动 VC。

图 1-1 VC++启动

图 1-2 VC++图标

1.1.3 开始一个新程序

一、建立工程

启动 VC 环境后，选择"File"菜单中的"New"项，弹出图 1-3 所示的对话框。其标签项自动选择为"Projects"（工程）。作为初学者，在左侧的列表中选择倒数第 3 项"Win32 Console

Application"（win32 控制台应用），在右侧"Project Name"文本框中输入欲建立的工程名称，如：Example01。在"Location"（位置）中选择工程的存放位置（最好为课程建一个总文件夹），单击"OK"按钮，系统出现建立工程的导航对话框，如图1-4所示。

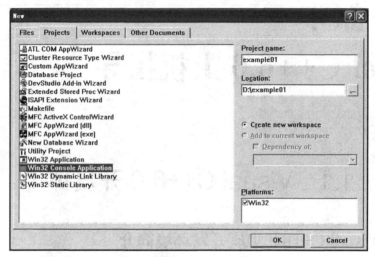

图1-3 创建控制台应用

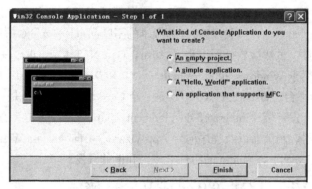

图1-4 创建工程的导航对话框

在图1-4的导航对话框中选择默认的"An empty project"（空的工程），然后单击"Finish"（完成）。系统弹出相关的创建工程的信息，如图1-5所示。单击"OK"按钮后，系统创建工程，建立相关的文件夹，不过这些文件夹都是空的，系统界面类似于图1-6。

图1-5 创建工程的信息

在图1-6中，窗口左侧为"Workerspace"(工作空间)窗口，下部窗口为"Output"（输出）窗口。在workerspace（工作空间）窗口中有两个视图标签：ClassView（类视图）和FileView（文件视图）。前者按照C++"类"的管理方式展现C/C++的源代码，后者按照文件的组织方式展现C/C++的源代码。单击标签的名称，可以在二者之间切换显示。

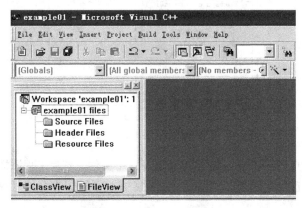

图1-6 创建工程后的界面

接下来的操作是为工程增加一个源文件，具体步骤见添加源程序部分内容。

二、添加源程序

1. 源文件的添加

按照方法一建立工程后，整个工程是一个完全空的架子，没有任何源文件。此时，单击菜单"File"，再次选择"New"，系统弹出与图1-3一样的对话框，不过，此时的对话框默认的标签是"Files"，如图1-7所示。

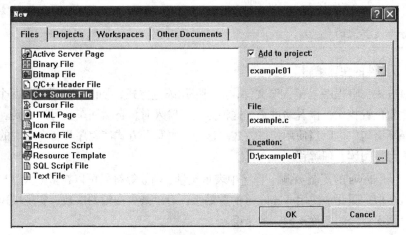

图1-7 为工程建立源文件的对话框

在图1-7中，选择"C++Source Files"（C++源文件），在右边的File文本框中填写文件名（一定要带有扩展名c或cpp，它们的编译器不同），然后单击"OK"，系统为使用者建立源文件并打开，右边的文本编辑区，就是展示源文件内容的窗口。图1-8是为工程example01添加了源文件exam01.c后，并展开了Source Files文件夹（单击其前面的"+"号）后的界面。

图 1-8 工程添加源文件 exam01.c 后的界面

双击文件名 exam01.c，可以在右边的窗口打开这个文件。若为工程添加了多个文件，则每个打开的文件对应一个窗口，这个窗口代表源文件的文本区，可以在此编写程序。例如，编辑如下代码：

```
/* 程序 1 */
#include "stdio.h"
int main()
{
    printf("这是第一个c程序!\n");
    printf("hello world!\n");
    return 0;
}
```

输入汉字后要及时切换回到英文输入方式，因为 C 语言使用的\n、引号、分号、括号等都是英文的。

2. 程序源文件的快速编辑

采用缩进方式：将程序 1 键入源文件，应当采用缩进方式，如图 1-9 所示的两个 printf 行，与大括号的垂直位置相比，位置缩进了。这种方式在输入时，自然回车就可以做到。以后的编程中，分支、循环语句中也应当使用缩进方式。这种"书写"方式使程序具备有层次的美感，增强逻辑感，让人容易阅读、理解程序。

积极使用 windows 的复制-粘贴，程序中表示变量、函数等符号可以在多处大量出现，甚至一些语句也极其类似，因此，使用 windows 的剪贴板，可以减少击键的次数，实现快速输入，还可以减少出现"两次键入的名称不同"的错误。

使用剪贴板，需要事先选定文本，其设计的方法如下。

光标定位：除了使用鼠标定位外，可以使用键盘上的光标移动键（箭头键）、home 键、end 键、快捷键等进行快速光标定位。

键盘右侧区域的箭头键←、→是水平（光标所在位置左右）移动一个字符。

键盘右侧区域的箭头键↑、↓是光标在垂直位置移动一行（上下行移动）。

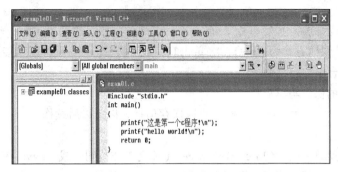

图 1-9　编写程序 1

文本选择：除了用鼠标拖动选择外，双击选择可以选择一个词汇（变量名、函数名等）；如果打算选择单行代码，可以在文本窗口左侧的边框上、该行的前面单击鼠标左键，即可完成单行的选择。如果打算选择多行，可在窗口左侧的边框上，按下鼠标左键进行拖动。

全文操作：全文选择可以使用键 Ctrl+A。

全文缩进：整个源程序按照格式进行缩进可以使用快捷键 Alt+F8，当然，必须事先选择全文。

取消与恢复：前次的操作可以用快捷键 Ctrl+Y 取消；而刚刚被取消的操作可以用快捷键 Ctrl+Z 恢复。

三、编译与运行

1. 编译

当将程序 1 键入之后，可单击菜单"组建"中的"编译"，系统会对源文件及整个工程进行编译，编译完成后再进行连接，最终生成可执行程序（exe）；也可以单击菜单第 2 项或按 F7 生成可执行程序，如图 1-10 所示。

系统进入编译时，在下面将出现 output 窗口，其中给出编译和链接过程中的语法检查信息。如果有错，给出错误信息。这些信息包括错误行、类别、错误代号、错误对象名、错误原因。

图 1-10　编译与连接

2. 运行程序

在图 1-10 所示的组建菜单中，有一个带有叹号图标的菜单项——Execute（执行）example01.exe，（其快捷方式为 Ctrl+F5），单击它，或者按 Ctrl+F5 就可以直接运行程序。图 1-11 就是运行时的控制台窗口。

四、工程项目的保存与再启动

当工程项目没有完成而需要暂停时，可以保存项目，以便以后续作。

1. 保存工程

单击"文件"菜单中的"Close Workspace"（关闭工作空间）或"Save Workspace"（保存工作空间）都可以保存工程，前者在没有保存时，给出提示保存。

注意：只有关闭工作空间以后，才能编写下一个程序，如图 1-12 所示。

图 1-11　运行结果　　　　　　　　图 1-12　关闭工作空间

2. 打开工程

单击"文件"菜单中的"Open Workspace"（打开工作空间），在出现的对话框中，选择正确的工程文件夹，后打开后缀为.dsw 的文件。

另一个打开已有工程的方法是，单击"文件"菜单中的"Recent Workspaces"（最近的工作空间），从列表中选择。

第 3 个方法是在 windows 资源管理器中，找到并打开相应的工程文件夹后，双击后缀为.dsw 的文件。

用打开工程的方法打开工程，工程的状态为上次操作的最后状态，打开后就可以接着操作。

1.2　C 语言程序调试方法

1.2.1　错误的类型

语法错误：拼写错，括号不匹配，漏写分号，等等。对于查出的错误（Error）必须排除，否则程序无法运行；而警告（Warning）则应根据情况处理，否则可能产生运算误差等。

逻辑错误：编译无误，有时也有执行结果，但结果不符合题意。例如 scanf 的参数中漏写地址符，if 语句、for 语句花括号位置错误等，都会导致此类错误。

运行错误：运行结果错误也可能是由于输入数据错误、类型不匹配等造成的。例如用户没有按照 scanf 规定的格式输入数据就会造成此类错误。

1.2.2 查错的方法

静态检查：人工检查，程序的结构、各函数间的调用关系，拼写检查。

编译程序：由 C 编译系统对程序进行查错，根据错误提示找出错误的位置并改正。注意提示的出错行未必是真正出错的行，常需要向上面的行寻找；而且系统指出的错误类型也未必是真正的错误，需要分析，不能停留在字面上。代码中有一个错误时，可能产生一大批编译错误，应从上到下逐一改正，修改一两个后再次编译。

排除语法错误后，运行程序，输入数据，得出结果，还应对结果进行分析，看是否符合要求。要准备一些测试数据，有意识地检查结果的正误。

若运行结果错误，通常由于程序中存在逻辑错误，应对照流程图检查算法逻辑。

对于怀疑出错的地方，添加一些 printf 函数输出某些变量的值，以找到出错的程序段，缩小查错范围。

1.2.3 Visual C++ 6.0 中常用的程序调试工具

Visual C++ 6.0 提供单步运行、断点跟踪等工具，帮助程序员查错。

一、设置固定断点或临时断点

所谓断点，是指定程序中的某一行，让程序运行至该行后暂停运行，使得程序员可以观察分析程序的运行过程中的情况。这些情况一般包括以下几个方面。

（1）在变量窗口(Varibles)中观察程序中变量的当前值。程序员观察这些值的目的是与预期值对比，若与预期值不一致，则此断点前运行的程序肯定在某个地方有问题，以此可缩小故障范围。

（2）在监控窗口（Watch）中观察指定变量或表达式的值。当变量较多时，使用 Varibles 窗口可能不太方便，使用 Watch 窗口则可以有目的、有计划地观察关键变量的变化。

（3）在输出窗口中观察程序当前的输出与预期是否一致。同样地，若不一致，则此断点前运行的程序肯定在某个地方有问题。

（4）在内存窗口（Memory）中观察内存中数据的变化。在该窗口中能直接查询和修改任意地址的数据。对初学者来说，通过它能更深刻地理解各种变量、数组和结构等是如何占用内存的，以及数组越界的过程。

（5）在调用堆栈窗口（Call Stack）中观察函数调用的嵌套情况。此窗口在函数调用关系比较复杂或递归调用的情况下，对分析故障很有帮助。

二、使用功能键或相应的菜单项进行操作。

F9：在当前光标所在行设置断点（Breakpoint，再次使用则取消当前行已有断点），见图 1-13。

F5（Go）：调试状态运行程序，程序执行到有断点的地方停下，见图 1-14。

此时左下角的变量窗口（Variables）显示当前的变量值。

如果在右下角的察看窗口（Watch）输入变量名，则可监控该变量值的变化。

F10（Step Over）或 F11（Step Into）：单步执行程序。

F11 和 F10 的区别：如果当前执行语句是函数调用，则 F11 会进入被调用函数里面。

Ctrl+F10：运行到光标所在行。

Shift+F11：跳出当前所在函数。

Shift + F5（Stop Debugging）：停止调止状态。

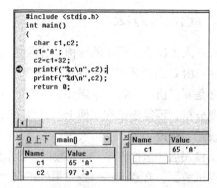

图 1-13 断点　　　　　　　　图 1-14 运行到断点处，变量窗口，察看窗口

在调试状态，系统会出现 Debug（调试）工具箱，如图 1-15 所示。其中包括了上面提到的功能的工具按钮。

图 1-15 调试工具箱

三、单步运行调试的基本步骤

（1）保存 C 或 C++ 文件。

（2）根据断点调试找到错误处。

（3）采用 F10 或 F11 单步调试找到精确的错误处。可先用 F10，确定函数输入、输出是否与预想的一致；如不相符，则用 F11 进入函数体一步一步调试。

（4）调试过程中，需要监视程序中变量值的变化。VC++ 6.0 的 Variables 和 Watch 窗口，就用来设置监视变量。在调试过程中，鼠标轻轻放在变量上（不用单击）也会显示该变量的值。

1.3　C 语言的编程风格

遵循一门语言的编程风格是非常重要的，否则编写的代码将难以阅读，给后期的维护带来诸多不便，例如，一个程序员将许多代码都写在同一行，尽管程序可以正确地编译和运行，但是这样的代码几乎无法阅读，其他程序员无法容忍这样的代码，造成沟通困难。本节将介绍一些最基本的编程风格，希望初学者在关注编程的正确性同时注重自己编写程序的可读性。

在编写程序时，许多地方如函数体、循环语句的循环体及分支语句的分支体都涉及使用一对大括号括起来的多行语句，俗称"代码块"。"代码块"有两种流行（行业内遵循的习惯）的写法——Allmans 风格和 Kernighan 风格。下面分别介绍这两种风格。

1.3.1　编程风格

Allmans 风格也称"独行"风格，即左右大括号各自独占一行，如下列代码所示。

```
#include<stdio.h>
int main()
{
```

```
    printf("Programing");
    printf("is fun.\n");
    return 0;
}
```

当代码量较小时适合使用 Allmans 风格，代码布局清晰，可读性强，本书的大部分例题都采用了这个风格。

Kernighan 风格也称"行尾"风格，既左大括号在上一行的行尾，而右大括号独占一行，如下列代码所示。

```
#include<stdio.h>
/*getMax 函数用于求两个整数中的最大值，
其中参数 a，b 的类型和返回值类型都是整型的。      */
int getMax(int a,int b){
      if(a>b)
      return a;
      return b;
}
int main(){
  int x,y,max;
  printf("Please input two integers: ");
  scanf("%d%d",&x,&y);       //通过键盘输入两个整数
  max=getMax(x,y) ;          //调用函数实现求两个整型数据的最大值
printf("The max number is:%d.\n",max);
  return 0;
}
```

当代码量较大时不适合使用 Allmans 风格，因为该风格将导致代码的左半部分出现大量的左右大括号，导致代码清晰度下降，这时应采用 Kernighan 风格。

1.3.2 变量名、函数名的命名规则

1. 变量名的命名规则

变量的命名规则要求用"匈牙利法则"。即开头字母用变量的类型，其余部分用变量的英文意思或其英文意思的缩写，尽量避免用中文的拼音，要求单词的第一个字母大写。

即变量名=变量类型+变量的英文意思（或缩写）对非通用的变量，在定义时加入注释说明，变量定义尽量可能放在函数的开始处。如：

short(int) 用 n 开头 nStepCount
long(LONG) 用 l 开头 lSum
char(CHAR) 用 c 开头 cCount
float(FLOAT) 用 f 开头 fAvg
double(DOUBLE) 用 d 开头 dDeta
void(VOID) 用 v 开头 vVariant

全局变量用 g_开头，如一个全局的长型变量定义为 g_lFailCount，即变量名=g_+变量类型+变量的英文意思（或缩写）。静态变量用 s_开头，如一个静态的指针变量定义为 s_plPerv_Inst，即变量名=s_+变量类型+变量的英文意思（或缩写）。对枚举类型（enum）中的变量，要求用枚举变量或其缩写做前缀，并且要求用大写。对结构体类型变量的命名要求定义的类型用大写，并要加上前缀，其内部变量的命名规则与变量命名规则一致。

2. 函数名的命名规则

函数的命名应该尽量用英文表达出函数完成的功能。遵循动宾结构的命名法则，函数名中动词在前，并在命名前加入函数的前缀，函数名的长度不得少于 8 个字母。一般第一个字母为小写英文，以后每个单词的第一个字母大写，例如：

```
long cmGetDeviceCount(……);
```

1.3.3 注释

注释是程序员对自己所编写程序的注释性说明，给代码添加合适的注释是一个良好的编程习惯。注释的目的是有利于代码的维护，提高代码可读性。C 语言支持两个格式的注释：单行注释和多行注释。注释的内容不限定语言，可以是中文也可以是英文。

单行注释使用//表示单行注释的开始，即该行中从//开始的后续内容都为注释。如

```
    scanf("%d%d",&x,&y);        //通过键盘输入两个整数
        max=getMax(x,y) ;           //调用函数实现求两个整型数据的最大值
```

多行注释以/*表示注释开始，以*/表示注释结束，其中/*之间不能有空格。例如上例中：

```
#include<stdio.h>
/*getMax 函数用于求两个整数中的最大值,
其中参数 a,b 的类型和返回值类型都是整型的.       */
int getMax(int a,int b){
    if(a>b)
      return a;
    return b;
}
```

本书中，根据需要两种注释都会使用。

编译器在将程序翻译为机器语言代码时，会忽略所有的注释内容，也就是说，注释对程序功能没有任何影响。但注释是编写程序不可缺少的部分，清晰的注释记录了程序员编写程序时的思路和设计思想，也便于自己或其他程序员日后查阅、引用。

在上例中，要求用户输入两个整数前，会通过 printf("Please input two integers:"); 语句向屏幕上打印一行提示，非常直观。大家在编写程序时，也要注意编写程序的界面友好，输入输出要有提示说明，对用户错误的输入要有检测和提示，养成良好的编程习惯。

第 2 部分 基本实验

本部分为读者设置了 15 个实验项目，读者可根据需要选择。一些实验项目给出了参考程序，希望读者首先独立思考、独立编程，然后和书中给出的答案进行比较，考察两者的优劣。

一、实验要求

1. 实验前应明确实验目的，阅读实验内容，对于较复杂的题目给出算法的 N-S 图描述，写好实验题目程序清单。
2. 能够理解调试过程中出现的提示信息，尽量独立改正程序中的各类错误，提高自己调试程序的能力。
3. 随时记录调试程序时遇到的问题，养成良好的交流学风，在交流中找到解决问题的途径和方法。
4. 实验结束时要填写实验报告。
5. 充分利用上机时间，提高上机效率。上机前把需要运行的程序准备好，不要现场编写，因为上机的时间是用来调试程序的。

二、实验报告的内容

本部分设置的实验项目，主要包括以下几个部分。

1. 实验目的
2. 实验内容

简要写出要求验证的内容和要求设计的题目。

3. 实验步骤

对于设计性实验，给出算法说明（复杂的可用 N-S 图表示）和程序清单。

4. 实验结果

包括程序运行时数据的输入、运行结果的输出。

5. 实验总结

对运行情况以及本次实验所取得的经验进行分析。如调试时出现错误信息，对错误原因进行分析。

实验 1　初识 C 程序

一、实验目的

1. 编辑简单的 C 程序，以此熟悉 Visual C++ 6.0 的集成开发环境，并初步认识 C 程序的组成。

2. 熟悉 Visual C++ 6.0 对 C 程序的编辑、编译、连接和运行等步骤和方法。
3. 了解编译过程中的语法错误信息并了解修改错误的方法。
4. 初步掌握 C 程序的调试方法。

二、实验内容

1. 验证性实验

（1）按照下列要求完成相应的操作

① 启动 Visual C++6.0：单击"开始"|"程序"|"Microsoft Visual Studio 6.0"|"Microsoft Visual C++6.0"菜单项。

② 参照第 1 章内容，在 D 盘或其他盘上创建 C 源程序文件 exam_1.c，并在 Visual C++ 6.0 的代码窗口中输入和编辑以下程序代码（其中/* */为 C 语言的注释，注释可以不输入，每行前面的"行 1""行 2"等不输入，身高和体重只能输入整数）。

```
行1    #include <stdio.h>
行2    void main( )
       {
           int x , y , z ;     /* 定义的身高x、体重y均为整数 */
行3        printf("\n请输入你的身高（厘米）:" );
           scanf("%d ", &x ) ;
           printf("\n请输入你的体重（公斤）:" );
行4        scanf("%d " , &y);
           z = x - 105 ;       /*标准体重的计算公式*/
行5        if (z > y) printf("你比标准体重轻，要多吃多睡哦。\n ") ;
           if (z < y) printf("你超过了标准体重，要加强锻炼。\n ") ;
行6        if (z==y) printf("你就是标准体重，太牛了，继续保持!\n ") ;
            return 0;
       }
```

③ 用菜单命令或快捷键键对 exam_1.c 进行编译、连接，最终生成 exam_1.exe 文件，并找到该文件所在路径。

④ 按以下要求修改程序并编译，注意观察输出窗口中的错误信息，分析其原因，并对其进行修改。

删除行 1，然后重新编译。

删除行 2 中的"void"，然后重新编译。

删除行 3 中的最后一个分号，然后重新编译。

删除行 4 中的"&"，然后重新编译。

删除行 5 中的"\n"，然后重新编译，并连续运行两次。

修改行 6 中的"=="为"="，然后重新编译并运行。

⑤ 在源文件第一行前加上语句 #include "stdlib. h"，在函数体的最后一个大括号之前添加语句 system("cls ");，然后重修编译、连接并运行程序，观察结果。

⑥ 执行"工具"|"选项"，在"选项"对话框中，选择"目录"（directory）选项卡，分别查看"可执行文件"、"Include files"、"Library files"、"Source files"的路径。若选择"Include files"，用鼠标双击"路径"列表框中的"C:\Program Files\Microsoft Visual Studio\VC98\Include"，将盘符 C: 改为 D:，单击"确认"按钮，重新编译，观察出错信息并修改。

⑦ 采用单步跟踪和设置断点的方式调试程序，观察程序运行过程中变量值的变化。

（2）预测并写出下列程序的运行结果，然后上机验证预测结果是否正确。

```c
#include <stdio.h>
int main( )
{
printf("Hello world!\n ");
printf("-----\n ");
printf("-\n ");
printf("-\n ");
printf("-----\n ");
return 0;
}
```

（3）C 语言源程序文件的后缀是.C，经过编译之后，生成后缀为.OBJ 的机器语言目标文件，经连接生成后缀.EXE 的可执行文件。在桌面新建一个"C 程序"文件夹，新建一个 C 语言程序保存在上述文件夹中，运行如下代码，验证程序编译、连接、运行后是否生成上述文件。

功能：输入整数 a 和 b，交换 a 和 b 的值后输出。

```c
#include <stdio.h>
int main()
{
    int a,b,temp;
    printf("输入整数a,b:");
    scanf("%d%d",&a,&b);
    temp=a;
    a=b;
    b=temp;
    printf("a=%d  b=%d\n",a,b);
    return 0;
}
```

2. 设计性实验

（1）模仿以下程序，编写一个简单乘法器，输入任意两个实数，程序输出它们的乘积。在 vc 中编辑、调试和运行，并分析结果是否正确。

```c
#include <stdio.h>
int main( )
{
float x,y,result ;
printf("请输入两个实数,以空格隔开: ");
scanf("%f %f ", &x,&y);
result = x + y ;
printf("%f + %f = %f \n " ,x,y,result);
return 0;
}
```

（2）编写程序解决以下问题。已知飞机由上海于 6 月 2 日 17 时飞往美国旧金山，飞行时间为 14 小时，计算飞机到达目的地的当地时间。

提示

上海在东 8 区，旧金山在西 8 区，东早西晚。

三、实验注意事项

1. C 程序运行必须从 main 函数开始，因此一个 C 程序要有一个 main 函数，且只能有一个 main 函数。当一个程序运行结束之后，一定要先选择"文件"->"关闭工作空间"命令关闭工作空间，然后再开始创建一个新的 C 程序。

2. 程序输入过程中应注意的问题。

（1）区分大小写。

（2）所有的符号都是英文状态下。

（3）程序中需要空格的地方一定要留空格。

实验 2　C 语言基础

一、实验目的

1. 掌握 C 语言中的数据类型，不同数据类型的定义与表示范围。
2. 掌握 C 语言中各种不同运算符的使用方法。
3. 掌握由运算符组成的表达式及表达式中不同数据类型的转换原则。
4. 掌握简单 C 程序的编写方法。

二、实验内容

1. 验证性实验

（1）要使下面程序的输出语句在屏幕上显示：1,3,57，则从键盘输入的数据格式应为以下备选答案中的_____。

```
#include <stdio.h>
int main()
{
    char a,b;
    int c;
    scanf("%c%c%d",&a,&b,&c);
    printf("%c,%c,%d\n",a,b,c);
    return 0;
}
```

A. 1 3 57　　　　B. 1,3,57　　　　C. '1','3',57　　　　D. 1 3 57

思考 1　在与上面程序的键盘输入相同的情况下，要使上面程序的输出语句在屏幕上显示为：1 3 57，则应修改程序中的哪条语句，怎样修改？

思考 2　要使上面程序的键盘输入数据格式为：1,3,57，输出语句在屏幕上显示的结果也为：1,3,57，则应修改程序中的哪条语句，怎样修改？

思考 3　要使上面程序的键盘输入数据格式为：1,3,57，而输出语句在屏幕上显示的结果为：'1','3',57，则应修改程序中的哪条语句，怎样修改？

利用转义字符输出字符单引号字符。

（2）执行下列程序后，其输出结果是（ ）。

```
#include <stdio.h>
int main()
{
int  a=9;
a+=a-=a+a;
printf("%d\n",a);
return 0;
 }
```

A．18　　　B．9　　　C．-18　　　D．-9

注意此表达式的运算顺序是从右向左。

（3）假设所有变量均为整型，表达式:a=2，b=5，a>b?a++:b++，a+b 的值是（ ）。
A．7　　　B．8　　　C．9　　　D．2

?:是 C 语言中唯一的三目运算符，注意其运算规则，可以实现简单的选择结构。

（4）输入如下源程序，然后进行编译、连接和运行。

```
#include <stdio.h>
int main()
{
  int i,j,x,y;
  i=3;
  j=5;
  x=i++;
  y=j++;
  printf("i=%d,j=%d\n ",i,j);
  printf("x=%d,y=%d\n ",x,y);
  i=3;
  j=5;
  printf("i=%d,j=%d\n ",++i,++j);
  return 0;
}
```

运行程序，观察输出的结果，注意屏幕上两次输出的 i、j、x、y 各个变量的值，体会前置运算和后置运算的差异。

（5）能正确表示逻辑关系"a≥10 或 a≤0"的 C 语言表达式是（ ）。
A．a>=10 or a<=0　　B．a>=0|a<=10　　C．a>=10 && a<=0　　D．a>=10 || a<=0

C语言中的表达式和数学表达式的区别。

（6）已知 i=5，写出语句 a=(a=i+1, a+2, a+3); 执行后整型变量 a 的值是_____。

　　　　逗号表达式的结果为最后一个表达式的值，本题为 a+3 的值，建议编写代码验证。

（7）以下程序的输出结果是（　）。
```
#include <stdio.h>
int main()
{
float x=3.6;
   int  i;
   i=(int)x;
 printf("x=%f,i=%d\n",x,i);
return 0;
}
```
A. x=3.600000, i =4   B. x=3, i=3   C. x=3.600000, i=3   D. x=3 i=3.600000

　　　　变量的强制转换并不改变原变量的值，只是将转换的值放在一个中间变量中。

（8）执行下列语句后，a 和 b 的值分别为（　）。
```
int a,b; a=1+'a';
b=2+7%-4-'A';
```
A. -63, -64   B. 98, -60   C. 1, -60   D. 79, 78

　　　　①'a'可以转化为ASCII值97，然后进行运算。②%运算符的结果符号和被除数的符号一致。

（9）编写代码计算表达式 1.234&&5.982 的值是_____。

　　　　逻辑表达式的结果只有两个：真和假。

（10）下列程序的输出结果为（　）。
```
#include <stdio.h>
int main()
{
int m=7,n=4;
   float  a=38.4,b=6.4,x;
   x=m/2+n*a/b+1/2;
   printf("%f\n",x);
return 0;
}
```

（11）执行下列程序段后，m 的值是_____。

16

```
#include <stdio.h>
int main()
{
   int w=2,x=3,y=4,z=5,m;
   m=(w<x)?w:x;
   m=(m<y)?m:y;
   m=(m<z)?m:z;
printf("%d",m);
return 0;
}
```

A. 4 　　　　B. 3 　　　　C. 5 　　　　D. 2

2. 设计性实验

（1）求 3 个数中的最大数。

提示　　　　可使用选择运算符。

（2）已知华氏温度转换为摄氏温度的公式为：$c = 5(f-32)/9$，其中 c 是摄氏温度，f 为华氏温度，编程实现将任意输入的华氏温度值转换为摄氏温度值输出。

```
#include <stdio.h>
int main()
{
   int f;
float c;
   scanf("%d",&f);
   c = 5.0 / 9.0 * (f - 32);
   printf("%.2f",c);
   return 0;
}
```

三、实验注意事项

1. 要注意变量的数据类型、输入语句和输出语句中使用的格式字符应与数据类型相对应。
2. 使用 getchar 和 putchar 函数，应在文件开头加上语句#include "stdio.h"；使用数学函数，应在文件开头加上#include "math.h"。

实验 3　顺序结构程序设计

一、实验目的

1. 掌握 C 语言中的赋值语句的使用方法。
2. 掌握 C 语言中基本输入/输出函数的调用方法。
3. 掌握各种类型数据的输入/输出方法，能正确使用各种格式符。
4. 掌握预处理命令#define、#include 的使用方法。

二、实验内容

1. 验证性实验

（1）已知圆柱体横截面圆半径 r，圆柱高 h。编写程序，计算圆周长 l、圆面积 s 和圆柱体体积 v，并输出计算结果。

```c
/* 计算圆周长、面积及圆柱体体积的程序 */
#include <stdio.h>
#define PI 3.14159
int main()
{
  float r,h,l,s,v;
  printf("r,h=");
  scanf("%f,%f",&r,&h);
  l=2*PI*r;
  s=PI*r*r;
  v=s*h;
  printf("L=%f\tS=%f\tV=%f\n",l,s,v);
  return 0;
}
```

（2）设银行定期存款的年利率为 rate，并已知存款期为 n 年，存款本金为 capital 元，试编程计算 n 年后的本利之和 deposit。要求定期存款的年利率 rate、存款期 n 和存款本金 capital 均由键盘输入。

```c
#include <math.h>
#include <stdio.h>
int main()                                              /*主函数首部*/
{
int n;                                                  /*存款期变量声明*/
   double rate;                                         /*存款年利率变量声明*/
   double capital;                                      /*存款本金变量声明*/
   double deposit;                                      /*本利之和变量声明*/
   printf("Please enter rate, year, capital:");         /*打印用户输入的提示信息*/
   scanf("%lf,%d,%lf", &rate, &n, &capital);            /*输入数据*/
   deposit = capital * pow(1+rate, n);                  /*计算存款利率之和，pow 为幂函数*/
   printf("deposit = %f\n", deposit);                   /*打印存款利率之和*/
   return 0;
}
```

程序调试时要注意如下几点。

① 输入数据的格式要与程序中要求的格式一致。如上述程序要用","分隔数据。
② 根据程序运行情况，调整输入、输出数据的格式，使数据的输入/输出格式更符合使用习惯。
③ 运行程序，输入负数数据，查看程序的执行结果。

（3）从键盘输入一个大写字母，要求改用小写字母输出。

```c
#include <stdio.h>
int main()
{
  char c1, c2;
  printf("Please input a char:");
  c1=getchar();
  printf("%c,%d\n",c1,c1);
```

```
    c2=c1+32;
    printf("%c,%d\n",c2,c2);
    return 0;
}
```

（4）以下程序的输出结果是（　）。

```
#include <stdio.h>
int main()
{
int  i,j,k,a=3,b=2;
 i=(--a==b++)?--a:++b;
 j=a++;
k=b;
 printf("i=%d,j=%d,k=%d\n",i,j,k);
return 0;
}
```

A．i=2, j=1, k=3　　　B．i=1, j=1, k=2　C．i=4, j=2, k=4　　D．i=1, j=1, k=3

注意变量赋值的位置。

（5）编程验证 int，float，double，char 四种数据类型的存储长度。

```
#include <stdio.h>
int main()
{
char  x=65;
float  y=7.3;
int  a=100;
double  b=4.5;
printf("%d %d\n",sizeof(char),sizeof(x));
printf("%d %d\n",sizeof(float),sizeof(y));
printf("%d %d",sizeof(int),sizeof(a));
printf("%d %d",sizeof(double),sizeof(b));
return 0;
}
```

使用长度运算符 sizeof

（6）根据输入的整型变量 m 的值，计算如下公式的值：$y=\sin(m) \times 10$。

例如：若 m=9，则应输出：4.121185。

```
#include "stdio.h "
#include "math.h "
int main()
{
    int m;
    float n;
    printf("Enter m: ");
    scanf("%d", &m);
    n=sin(m)*10;
    printf("\nThe result is %1f\n",n);
    return 0;
}
```

（7）从键盘输入所需数据，求出 z 的值并输出，要求输出结果保留 2 位小数。补全下面的程序代码。

```
#include <stdio.h>
_____
int main()
{
int x;
double y,z;
    scanf("____",&x,&y);
    z=2*x*sqrt(y);
    printf("z=____",z);
    return 0 ;
}
```

运行程序，查看结果是否保留了 2 位小数。

（8）从键盘上输入两个复数的实部与虚部，求出并输出它们的和、差、积、商。补全下面的程序代码。

```
#include<stdio.h>
int main()
{
    float a,b,c,d,e,f;
    printf("输入第一个复数的实部与虚部：");
    scanf("%f, %f",&a,&b);
    printf("输入第二个复数的实部与虚部：");
    scanf("%f, %f",&c,&d);
    _____;
    f=b+d;
    printf("相加后复数：实部：%f,虚部：%f\n",e,f);
    e=a*c-b*d;
    _____;
    printf("相乘后复数：实部：%f,虚部：%f\n",e,f);
    e=(a*c+b*d)/(c*c+d*d);
    _____;
    printf("相除后复数：实部：%f,虚部：%f\n",e,f);
    return 0;
}
```

2. 改错性实验

（1）求两数的平方根之差。

```
#include "stdio.h "
int main()
{
    double a,b,y;
    printf("Please enter two integer: ");
    scanf("%f%f ",&a, &B);
    y=sqrt(a)-sqrt(B);
    printf("y=%f\n ",y);
    return 0 ;
}
```

（2）输入圆的半径值，求二分之一的圆面积。例如：输入圆的半径值：19.527 输出为：s = 598.950017。

```
#include "stdio.h "
int main()
{
   float x;
   double s;
   printf ( "Enter x: ");
   scanf ( "%d", &x );
   s=1/2*3.14159* r * r;
   printf (" s = %f\n ", s );
return 0;
}
```

（3）已知圆锥半径 r 和高 h，计算圆锥体积 v。请纠正程序中存在错误。

```
#include <stdio.h>
int main()
{
float r=10,h=5;
v=1/3*3.14159*r2*h;
printf("v=%d\n",v);
return 0;
}
```

（4）按下列公式计算并输出 x 的值。其中 a 和 b 的值由键盘输入。请纠正程序中存在的错误，使程序实现其功能。

公式：$x=2ab/(a+b)^2$

```
#include <stdio.h>
 int main()
{
   int a,b;
   double x;
   scanf("%d,%d",a,b);
   x=2ab/(a+b)(a+b);
   printf("x=%d\n",x);
   return 0;
}
```

3. 设计性实验

（1）编写程序，从键盘随机输入 a、b 两个正整数，输出 a 除以 b 得到的商和余数。

```
#include <stdio.h>
 int main()
{
   int  a,b,s,y;
   printf("please input two integers: ");
   scanf("%d%d ",&a,&b);
   s=a/b;
   y=a%b;
   printf("a/b=%d\n ",s);
   printf("a%%b=%d\n ",y);
return 0;
}
```

（2）编写程序，实现如下功能：从键盘上输入某学生 4 门课程的成绩，计算该学生的总成绩和平均成绩。

```
#include <stdio.h>
 int main()
{
    float  m,c,h,r,sum,aver;
    printf("Please input four scores: ");
    scanf("%f%f%f%f ",&m,&c,&h,&r);
    sum=m+c+h+r;
    aver=sum/4;
    printf("The total is %f, average score is %f ",sum,aver);
    return 0;
}
```

（3）假设 *m* 是一个三位数，编写程序输出由 *m* 的个位、十位、百位反序而成的三位数（例如：123 反序为 321）。

```
#include <stdio.h>
 int main()
{
    int m,n,a,b;
    scanf("%d",&a);
    m=a/100;
    n=(a-m*100)/10;
    b=a%10;
    k=b*100+n*10+m;
    printf("k=%d",k);
    return 0;
}
```

三、实验注意事项

1. 输入函数 scanf()中输入变量前面不能漏掉&符号。
2. 输入多项数据时，注意间隔符号。

实验 4　选择结构程序设计

一、实验目的

1. 熟悉关系表达式和逻辑表达式的使用。
2. 掌握 if 语句、switch/case 语句的用法及其条件表达式。
3. 掌握 break 语句的用法。
4. 学会跟踪调试程序，针对具体程序及其相应的测试数据，观察程序运行是否达到预期结果。

二、实验内容

1. 验证性实验

（1）下列程序的功能是输入一个正整数，判断是否能被 3 或 7 整除，若不能被 3 或 7 整除，就输出"YES"，否则就输出"NO"。

```
#include <stdio.h>
int main( )
{
int k;
printf("请输入一个正整数:\n ");
scanf ("%d",【?】);
if (【?】)
printf("YES\n");
else
printf ("NO\n");
   return 0;
   }
```

（2）以下程序是实现输出 x, y, z 三个数中的最大者。

```
#include <stdio.h>
int main()
{
  int x = 4, y = 6,z = 7;
   int u ,【?】;
    if(x>y)
       【?】;
    else  u = y;
    if(u>z)
       v = u;
    else
       v= z;
    printf("the max is %d",v );
    return 0;
}
```

（3）计算一元二次方程的根。

```
#include <stdio.h>
#include 【?】
int main()
{
  double x1,x2,imagpart;
   float a,b,c,disc,realpart;
   scanf("%f%f%f",&a,&b,&c);
   printf("the equation");

  if(【?】<=1e-6)
     printf("is not quadratic\n");
   else
     disc=b*b-4*a*c;
   if(fabs(disc)<=1e-6)
     printf("has two equal roots:%-8.4f\n",-b/(2*a));
   else if(【?】)
   {
    x1=(-b+sqrt(disc))/(2*a);
    x2=(-b-sqrt(disc))/(2*a);
    printf("has distinct real roots:%8.4f and %.4f\n",x1,x2);
   }
   else
   {
    realpart=-b/(2*a);
```

```
        imagpart=sqrt(-disc)/(2*a);
        printf("has complex roots:\n");
        printf("%8.4f=%.4fi\n",realpart,imagpart);
        printf("%8.4f-%.4fi\n",realpart,imagpart);
    }
    return 0;
}
```

（4）运行 5 次下列程序，输入的数据分别是 1、2、3、4、5，观察每次的运行结果，并按照单步跟踪的方法和设置断点的方法分析其结果。

```
#include <stdio.h>
int main()
{
    int  n, a = 0, c = 0, d = 0, k = 0 ;
    printf("请分次输入数据（1/2/3/4/5）:\n ");
    scanf("%d ", & n );
    switch( n )
    {
        default : a ++ ;
        case 1: b ++ ; break ;
        case 2: c ++ ;
        case 3: d ++ ; break ;
        case 4: k ++ ;
    }
    printf(" a= %d, b= %d, c= %d, d= %d, n= %d " , a , b , c , d , n );
    return 0;
}
```

（5）下列程序是根据菜单功能计算圆、矩形或三角形的面积。请填空后调试程序，使其运行结果如图 2-1 所示。

图 2-1 计算图形面积

```
#include <stdio.h>
#include <math.h>
#define PI 3.141592654
int main()
{
    int choice;
    double r,a,b,c,sanjiao,s;
    printf("%5s*****计算图形面积的功能菜单 *****\n", " ");
    printf("%10s 1 - 圆面积\n", " ");
    printf("%10s 2 - 矩形面积\n", " ");
    printf("%10s 3 - 三角形面积\n", " ");
    printf("%10s 4 - 输出提示信息\n", " ");
    printf("请输入选择（1/2/3/4）: "); /*提示信息*/
    _____;
    switch(choice)
```

```
        {
            case 1:  //计算圆面积
                printf("请输入圆的半径:\n");
                scanf(" %lf ", &r);
                printf("圆的面积是%.3f \n",PI * r * r );
                _____;
            case 2:  //计算矩形面积
                printf("请输入矩形的长和宽(之间用逗号隔开):\n");
                scanf("%lf,%lf", &a , &B. ;
                printf("矩形的面积是%.3f \n", a * b );
                break;
            case 3:  //计算三角形面积
                printf("请输入三角形的三个边长(之间用逗号隔开):\n");
                scanf(" %lf,%lf,%lf", &a, &b, &c);
                if( _____ )  //判断是否是三角形
                {
                    s=(a+b+c)/2;
                    sanjiao=sqrt(s*(s-a)*(s-b)*(s-c));
                    printf("三角形的面积是%.3f \n", sanjiao);
                }
                else
                    printf("不能构成三角形" );
                break;
            default: printf("选了无效的菜单项\n");
        }
        return 0;
}
```

（6）输入某年某月某日，判断这一天是这一年的第几天？

```
#include <stdio.h>
int main()
{
  int day,month,year,sum,leap;
  printf("\nplease input year,month,day\n");
  scanf("%d,%d,%d",&year,&month,&day);
  switch(month)
  {
   case 1:sum=0;break;
   case 2:sum=31;break;
   case 3:sum=59;break;
   /***********SPACE***********/
   case 4:【?】;break;
   case 5:sum=120;break;
   case 6:sum=151;break;
   case 7:sum=181;break;
   case 8:sum=212;break;
   case 9:sum=243;break;
   case 10:sum=273;break;
   case 11:sum=304;break;
   case 12:sum=334;break;
   default:printf("data error");break;
  }
  /***********SPACE***********/
  【?】;
```

```
        /***********SPACE***********/
        if(year%400==0||(【?】))
           leap=1;
        else
           leap=0;
        /***********SPACE***********/
        if(【?】)
           sum++;
        printf("it is the %dth day.",sum);
        return 0;
}
```

（7）将字母转换成密码，转换规则是将当前字母变成其后的第四个字母，但 W 变成 A，X 变成 B，Y 变成 C，Z 变成 D。小写字母的转换规则同样。

```
#include <stdio.h>
int main()
{
   char c;
   /***********SPACE***********/
   while((c=【?】)!='\n')
   {
      /***********SPACE***********/
      if((c>='a'&&c<='z')||(c>='A'&&c<='Z'))【?】;
      /***********SPACE***********/
      if((c>'Z'【?】c<='Z'+4)||c>'z')  c-=26;
      printf("%c",c);
   }
   return 0;
}
```

2. 改错性实验

（1）下面程序的功能是求解函数 $y = \begin{cases} 2x, & x < 1 \text{ 或 } x > 10 \\ x^2, & 1 \leq x \leq 10 \end{cases}$

修改程序，使程序的功能与要求相符合。

```
#include <stdio.h>
int main()
{
    float x , y ;
    printf("请输入 x 的值:");
    scanf("%f ", x) ;
    if ( x < 1 || x > 10) y = 2 * x ;
    if ( 1 <= x <= 10 ) y = x * x ;
    printf("y = %f " , y);
    return 0;
}
```

（2）输入两个实数，并按代数值由小到大输出。（输出的数据都保留 2 位小数）

```
#include <stdio.h>
int main()
 {
    float t
```

```
    float a, b ;
    scanf("%f %f",&a,&b);
    if(a<b)
    {
     t=a;
     a=b;
     b=t;
    }
    printf("%f ,%f\n",&a,&b);
    return 0;
}
```

（3）编写程序计算下列分段函数的值：

$$f(x)=\begin{cases} x\times x+x & x<0 且 x\neq -3 \\ x\times x+5x & 0\leqslant x<10 且 x\neq 2 及 x\neq 3 \\ x\times x+x-1 & 其他 \end{cases}$$

```
#include <stdio.h>
int main()
 {
  /**********FOUND**********/
  double y
  /**********FOUND**********/
  if (x<0 ||x!=-3.0)
    y=x*x+x;
  else if(x>=0 && x<10.0 && x!=2.0 && x!=3.0)
    y=x*x+5*x;
  else
    y=x*x+x-1;
  /**********FOUND**********/
  printf("x=%f,f(x)=%f\n",x,y);
  return 0;
}
```

（4）下列程序功能：给定一个不多于 5 位的正整数，求它是几位数，并逆序打印出各位数字。

```
#include <stdio.h>
int main( )
{
  /**********FOUND**********/
  long a,b,c,d,e,x,
  scanf("%ld",&x);
  a=x/10000;
  /**********FOUND**********/
  b=x/10000/1000;
  c=x%1000/100;
  d=x%100/10;
  e=x%10;
  /**********FOUND**********/
  if (a==0)
    printf("there are 5, %ld %ld %ld %ld %ld\n",e,d,c,b,a);
  else if (b!=0)
    printf("there are 4, %ld %ld %ld %ld\n",e,d,c,b);
  else if (c!=0)
    printf(" there are 3,%ld %ld %ld\n",e,d,c);
  else if (d!=0)
```

```
       printf("there are 2, %ld %ld\n",e,d);
    else if (e!=0)
       printf(" there are 1,%ld\n",e);
    return 0;
}
```

（5）功能：编写一个程序模拟计算器的加、减、乘、除四则运算。

```
#include<stdio.h>
int main()
{
  float x,y;
  char operate1;
  printf("请输入 2 个数及操作数");
  scanf("%f%f ",x, y);
  switch(y)
    {
      case '+':
           x+=y;
           break;
      case '-':
           x-=y;
           break;
      case '*':
           x*=y;
           break;
      case '/':
           x/=y;
           break;
    }
  printf("%f",x);
  return 0;
}
```

（6）输入一行字符，分别统计出其中英文字母、空格、数字和其他字符的个数。

```
#include <stdio.h>
int main(){
  char c;
  int letters=0,space=0,digit=0,others=0;
  printf("please input some characters\n");
  /***********FOUND***********/
  while((c=getchar())=='\n')   {
  /***********FOUND***********/
    if(c>='a'&&c<='z'&&c>='A'&&c<='Z')
      letters++;
  /***********FOUND***********/
    else if(c=!' ')
      space++;
    else if(c>='0'&&c<='9')
      digit++;
    else
      others++;
  }
  printf("all in all:char=%d space=%d digit=%d others=%d\n",letters,
  space,digit,others);
  return 0;
}
```

（7）利用条件运算符的嵌套来完成此题：学习成绩 90 分或 90 分以上的同学用 A 表示，60～89 分的用 B 表示，60 分以下的用 C 表示。

```
#include <stdio.h>
int main()
{
  int score;
  /*********FOUND*********/
  char *grade;
  printf("please input a score\n");
  /*********FOUND*********/
  scanf("%d",score);
  /*********FOUND*********/
  grade=score>=90?'A';(score>=60?'B':'C');
  printf("%d belongs to %c",score,grade);
  return 0;
}
```

（8）一个 5 位数，判断它是不是回文数。即 12321 是回文数，个位与万位相同，十位与千位相同。

```
#include<stdio.h>
int main( )
{
  /*********FOUND*********/
  long ge,shi,qian;wan,x;
  scanf("%ld",&x);
  /*********FOUND*********/
  wan=x%10000;
  qian=x%10000/1000;
  shi=x%100/10;
  ge=x%10;
  /*********FOUND*********/
  if (ge==wan||shi==qian)
    printf("this number is a huiwen\n");
  else
    printf("this number is not a huiwen\n");
  return 0;
}
```

（9）编写一个程序计算某年某月有几天。（注意区分闰年）

```
#include<stdio.h>
int  main()
{
  int yy,mm,len;
  printf("year,month=");
  scanf("%d%d",&yy,&mm);
  /*********FOUND*********/
  switch(yy)
  {
    case 1:
    case 3:
    case 5:
    case 7:
    case 8:
    case 10:
    case 12:
```

```
            len=31;
            /**********FOUND**********/
            break
    case 4:
    case 6:
    case 9:
    case 11:
            len=30;
            break;
    case 2:
            if (yy%4==0 && yy%100!=0 || yy%400==0)
              len=29;
            else
              len=28;
            break;
    /**********FOUND**********/
    default
            printf("input error!\n");
            break;
   }
   printf("The length of %d %d id %d\n",yy,mm,len);
   return 0;
}
```

3. 设计性实验

（1）编写程序，判断任意一个 3 位正整数是否为回文数。回文数是指其各位数字左右对称的正整数，如 121。

（2）按以下公式计算自己的脂肪含量是否在正常范围内。

$$A = 腰围（in.）\times 4.15 \qquad （1\ in. = 2.54cm）$$
$$B = 体重（pt）\times 0.082 \qquad （1\ kg = 2.2pt）$$
$$脂肪含量 = (A - B - 76.76) \times 100\%$$

正常脂肪含量：男性 12%～20%，女性 20%～30%。

（3）根据学生成绩的等级输出相应的百分制成绩。成绩的等级 A、B、C、D、E 分别对应 90 分以上、80~89 分、70~79 分、60~69 分、60 分以下，使用 switch 语句编程实现。

（4）功能：从键盘上输入任意实数，求出其所对应的函数值。

$$z=e^x\ (x>10)$$
$$z=\lg(x+3)\ (x>-3)$$
$$z=\sin(x)/((\cos(x)+4)$$

三、实验注意事项

1. C 程序中表示比较运算的符号是 "=="，赋值运算符是 "="，不能混淆。

2. 控制表达式是指任意合法的 C 语言表达式（不只限于关系或逻辑表达式），只要表达式值为非零，则为 "真"，否则为 "假"。

3. 嵌套的 if 语句中，else 总是和它最近的尚未配对的 if 进行配对，可以利用{}改变配对。

4. case 及后面的常量表达式，仅仅起到语句标号的作用，要想完成选择分支，必须和 break 语句结合，在执行过程中，只要不遇到 break 语句，就一直往下执行，而不再判别是否匹配。实际应用中，允许多个 case 语句共用一个 break 语句。

实验 5　循环结构程序设计(一)

一、实验目的

1. 掌握循环语句中的 for、while 和 do...while 语句的基本使用方法。
2. 掌握循环条件的设定。
3. 理解不同循环的适用条件，如计数控制或条件控制循环，选择合适的循环语句，提高程序的效率。
4. 掌握常见问题的算法及其程序实现。

二、实验内容

1. 验证性实验

（1）键盘上输入 ABCdef 后回车，分析程序的运行结果。

```
#include <stdio.h>
int main()
{
  char ch;
  printf("please input a string\n");
  while((ch=getchar())!=\n)
  {
    if(ch>='A'&&ch<='Z')
      ch=ch+32;
    else if(ch>='a'&&ch<='z')
      ch=ch-32;
    printf("%c",ch);
  }
  printf("\n");
  return 0;
}
```

（2）下面程序的功能是求 1!+3!+5!+…+n!的和。完善程序并上机验证。

```
#include <stdio.h>
int main()
{
  long int f,s;
  int i,j,n;
  /***********SPACE***********/
  【?】;
  scanf("%d",&n);
  /***********SPACE***********/
  for(i=1;i<=n;【?】)
  {
    f=1;
    /***********SPACE***********/
    for(j=1;【?】;j++)
    /***********SPACE***********/
    【?】;
    s=s+f;
```

```
    }
    printf("n=%d,s=%ld\n",n,s);
    return 0;
}
```

（3）下列程序的功能是以每行5个数来输出300以内能被7或17整除的偶数，并求出其和。

```
#include <stdio.h>
int main()
{
  int i,n,sum;
  sum=0;
  /***********SPACE***********/
  【?】;
  /***********SPACE***********/
  for(i=1; 【?】 ;i++)
  /***********SPACE***********/
    if(【?】)
      if(i%2==0)
      {
        sum=sum+i;
        n++;
        printf("%6d",i);
        /***********SPACE***********/
        if(【?】)
          printf("\n");
      }
  printf("\ntotal=%d",sum);
  return 0;
}
```

（4）下列程序的功能为：输出100以内能被4整除且个位数为8的所有整数；请填写适当的符号或语句，使程序实现其功能。

```
#include <stdio.h>
  int  main()
    { int i,j;
/***********SPACE***********/
      for(i=0;【?】; i++)
        { j=i*10+8;
/***********SPACE***********/
          if (【?】)
/***********SPACE***********/
            【?】;
            printf("%d",j);
          }
      return 0 ;
    }
```

（5）程序功能：输出100到1000之间的各位数字之和能被15整除的所有数，输出时每10个一行。

```
#include <stdio.h>
int main()
{
  int m,n,k,i=0;
  for(m=100;m<=1000;m++)
```

```
    {
      /**********SPACE**********/
      【?】;
      n=m;
      do
      {
        /**********SPACE**********/
        k=k+【?】;
        n=n/10;
      }
      /**********SPACE**********/
      【?】;
      if (k%15==0)
      {
        printf("%5d",m);i++;
        /**********SPACE**********/
        if(i%10==0) 【?】;
      }
    }
    return 0;
}
```

（6）输出 9×9 口诀表。

```
#include <stdio.h>
int main()
{
    int i,j,result;
    printf("\n");
    /**********SPACE**********/
    for (i=1;【?】;i++)
    {
      /**********SPACE**********/
      for(j=1;j<10;【?】)
      {
        result=i*j;
        /**********SPACE**********/
        printf("%d*%d=%-3d",i,j,【?】);
      }
      printf("\n");
    }
    return 0;
}
```

（7）功能：编程求任意给定的 n 个数中的奇数的连乘积，偶数的平方和以及 0 的个数，n 通过 scanf()函数输入。

```
#include <stdio.h>
int main()
{
    int r=1,s=0,t=0,n,a,i;
    printf("n=");scanf("%d",&n);
    for(i=1;i<=n;i++)
    {
      printf("a=");
      /**********SPACE**********/
      scanf("%d",【?】);
      /**********SPACE**********/
```

```
        if(【?】!=0)
          /**********SPACE**********/
          【?】=a;
        else if(a!=0)
          /**********SPACE**********/
          s+=【?】;
        else
          t++;
    }
    printf("r=%d,s=%d,t=%d\n",r,s,t);
    return 0;
}
```

2. 改错性实验

（1）下面程序的功能是计算 n!。

```
#include <stdio.h>
int main()
{
    int i, n, s=1;
    printf("Please enter n: ");
    scanf("%d ",&n);
    for(i=1; i<= n; i++)   s = s * i;
    printf("%d! = %d\n ", n, s);
    return 0;
}
```

运行时先输入 *n*=4，输出结果为 4!=24，这是正确的。为了检验程序的正确性，再输入 *n*=10，输出为 10!=24320，这显然是错误的（正确值为 10!=3628800）。按 F10 键单步调试程序，同时在 VC 屏幕右下方的 watch 窗口添加变量 *i* 和 *s*，观察程序执行过程中变量 *i* 和 *s* 的值的变化，以找出程序的错误。分析原因，把程序改正过来，再用 *n*=20 进行实验，分析所得到的结果。

（2）程序功能：用下面的和式求圆周率的近似值。直到最后一项的绝对值小于等于 0.0001。

$$\frac{\pi}{4}=1-\frac{1}{3}+\frac{1}{5}-\frac{1}{7}+\cdots$$

```
#include <stdio.h>
/**********FOUND**********/
#include <stdlib.h>
int main()
{
  int i=1;
  /**********FOUND**********/
  int  s=0,t=1,p=1;
  /**********FOUND**********/
  while(fabs(t)<=1e-4)
  {
    s=s+t;
    p=-p;
    i=i+2;
    t=p/i;
  }
  /**********FOUND**********/
  printf("pi=%d\n",s*4);
  return 0;
}
```

（3）以下程序功能是输入一个整数 n，计算累加和(1+2+3+⋯+n)并输出。请找出错误并改正。
如输入 5↙

输出：The sum from 1 to 5 is 15

```
#include <stdio.h>
int main()
{
/**********FOUND**********/
    int i, n, sum;
/**********FOUND**********/
    scanf( "%d", n );
    for ( i = 0; i <= n; i++ );
/**********FOUND**********/
       sum += n;
    printf( "%d", sum );
return 0;
}
```

（4）功能：计算正整数 num 的各位上的数字之积。例如：输入 252，则输出应该是 20。

```
#include <stdio.h>
int main()
{ /**********FOUND**********/
long n;
long k;
  printf("\nPlease enter a number:");
 /**********FOUND**********/
  scanf("%ld", num);
  do
  {
    k*=num%10;
   /**********FOUND**********/
    num\=10;
  }while (num);
  printf("\n%ld\n",k);
  return 0;
}
```

（5）有 1、2、3、4 四个数字，能组成多少个互不相同且无重复数字的三位数？都是多少？

```
#include <stdio.h>
int main()
{
  int i,j,k;
 /**********FOUND**********/
  printf("\n")
 /**********FOUND**********/
  for(i=1;i<=5;i++)
    for(j=1;j<5;j++)
      for (k=1;k<5;k++)
      {
       /**********FOUND**********/
        if (i!=k||i!=j||j!=k)
          printf("%d,%d,%d\n",i,j,k);
      }
  return 0;
}
```

（6）一球从 100m 高度自由落下，每次落地后反跳回原高度的一半；再落下，求它在第 10 次落地时，共经过多少米？第 10 次反弹多高？

```
#include<stdio.h>
int main()
{
  /**********FOUND**********/
  float sn=100.0;hn=sn/2;
  int n;
  /**********FOUND**********/
  for(n=2;n<10;n++)
  {
    sn=sn+2*hn;
    /**********FOUND**********/
    hn=hn%2;
  }
  printf("the total of road is %f\n",sn);
  printf("the tenth is %f meter\n",hn);
  return 0;
}
```

（7）猴子吃桃问题：猴子第一天摘下若干个桃子，当即吃了一半，还不过瘾，又多吃了一个，第二天早上又将剩下的桃子吃掉一半，又多吃了一个。以后每天早上都吃了前一天剩下的一半零一个。到第 10 天早上想再吃时，见只剩下一个桃子了。求第一天共摘了多少。

```
#include <stdio.h>
int main()
{
  int day,x1,x2;
  day=9;
  /**********FOUND**********/
  x2==1;
  while(day>0)
  {
    /**********FOUND**********/
    x1=(x2+1)/2;
    x2=x1;
    /**********FOUND**********/
    day++;
  }
  printf("the total is %d\n",x1);
  return 0;
}
```

3. 设计性实验

（1）编写程序计算并输出：$1 + 12 + 123 + 1234 + \cdots$ 的前 n(设 0<n<10)项的和 sum，n 从键盘输入。例如输入 3，则输出 136；又如输入 6，则输出 137171。

（2）分别使用 while、for、do…while 语句计算 s=1!+3!+5!+7! +⋯+n!。

（3）求两个正整数 m 和 n 之间所有既不能被 3 整除也不能被 7 整除的整数之和。

（4）编程计算 1×2×3+3×4×5+⋯⋯+99×100×101。（正确结果 sum=13002450）

（5）输入 5 个整数 x，输出其中正整数的累加和 sum 与正整数的平均值 ave，输入输出格式如以下示例。

如输入：10 0 20 -5 31

则输出：Sum=61， Average=20.3

```
#include <stdio.h>
int main()
{
    int  i,x,n,sum;
/**********Program**********/



/********** End **********/
return 0;
}
```

三、实验注意事项

1. while、do…while、for 语句中的循环变量必须赋初值，正确设置循环条件，使循环有趋于终止的语句，否则有可能构成死循环。

2. while、do…while 语句什么情况下运行结果一致，什么情况下运行结果不同。

3. 注意{}的配对问题，正确使用{}构成复合语句。

实验 6　循环结构程序设计(二)

一、实验目的

1. 掌握循环语句中的 for、while 和 do…while 语句的特殊用法。
2. 掌握循环语句中的 for、while 和 do…while 语句实现循环嵌套的方法。

二、实验内容

1. 验证性实验

（1）读下列程序，分析其运行结果，然后编辑并运行，比较运行结果和分析结果是否一致。

```
#include <stdio.h>
int main()
{
    int i, j, s = 0;
    for(i = 0; i < 20; i++)
    {
        s = s + 1;
        for(j = 0; j < 3; j++)
        {
            if(j % 2) continue;
            s = s + 2;
        }
        if(s>10) break;
    }
```

```
            printf("s = %d, i= %d\n",s, i);
            return 0;
}
```

（2）下列程序的功能是输出由字符 w 构造成的形如 W 的图形，完善程序并调试。

```
 w              ww              w
  w            w  w            w
   w          w    w          w
    w        w      w        w
     ww              ww

#include <conio.h>
#include <stdio.h>
int main()
{
  int i,j,k,r,m;
  /***********SPACE***********/
  for(i=1; 【?】;i++)
  {
    for(j=1;j<=2;j++)
    {
      for(r=1;r<i;r++)printf(" ");
        printf("w");
      /***********SPACE***********/
        for(k=1; 【?】 ;k++)printf(" ");
          printf("w");
          for(m=1;m<i;m++)printf(" ");
    }
    /***********SPACE***********/
    【?】;
  }
  return 0 ;
}
```

（3）百马百担问题：有 100 匹马，驮 100 担货，大马驮三担，中马驮 2 担，两匹小马驮一担，求大、中、小马各多少匹？

```
#include <stdio.h>
int main()
{
  int hb,hm,hl,n=0;
  /***********SPACE***********/
  for(hb=0;hb<=100;hb+=【?】)
    /***********SPACE***********/
    for(hm=0;hm<=100-hb;hm+=【?】)
    {
      /***********SPACE***********/
      hl=100-hb-【?】;
      /***********SPACE***********/
      if(hb/3+hm/2+2*【?】==100)
      {
        n++;
        printf("hb=%d,hm=%d,hl=%d\n",hb/3,hm/2,2*hl);
      }
    }
  printf("n=%d\n",n);
```

```
    return 0 ;
}
```

（4）题目：计算 100~1000 有多少个数，其各位数字之和是 5。

```
#include<stdio.h>
int main()
{
        int i,s,k,count=0;
        for(i=100;i<1000;i++)
        {
                s=0;
                k=i;
/**********SPACE**********/
                while(【?】)
                {
                        s=s+k%10;
/**********SPACE**********/
                        k=【?】;
                }
                if(s!=5)
/**********SPACE**********/
                        【?】;
                else
                {
                        count++;
                        printf("%d %d\n",count,i);
                }
        }
        printf("个数为：%d\n",count);
        return 0;
}
```

（5）下列程序功能是求出一批非零整数中的偶数、奇数的平均值，用零作为终止标记。

```
#include <stdio.h>
int main()
{
  int x,i=0,j=0;
  float s1=0,s2=0,av1,av2;
  scanf("%d",&x);
/**********SPACE**********/
  while(x!=0)
  {
    if(x%2==0)
    {
      s1=s1+x;
      i++;
    }
/**********SPACE**********/
    【?】
    {
      s2=s2+x;
      j++;
    }
/**********SPACE**********/
    【?】;
  }
```

```
    if(i!=0)
       av1=s1/i;
    else
       av1=0;
    if(j!=0)
    /***********SPACE***********/
       【?】 ;
    else
       av2=0;
    printf("oushujunzhi:%7.2f,jishujunzhi:%7.2f\n",av1,av2);
    return 0;
}
```

（6）已知一个数列，它的头两项分别是 0 和 1，从第三项开始以后的每项都是其前两项之和。编程打印此数，直到某项的值超过 200 为止。

```
#include <stdio.h>
int main()
{
  int i,f1=0,f2=1;
  /***********SPACE***********/
  for(【?】;;i++)
  {
    printf("5%d",f1);
    /***********SPACE***********/
    if(f1>【?】) break;
    printf("5%d",f2);
    if(f2>200) break;
    if(i%2==0) printf("\n");
    f1+=f2;
    /***********SPACE***********/
    f2+=【?】;
  }
  printf("\n");
  return 0;
}
```

（7）编写程序，输出 1000 以内的所有完数及其因子。

说明：所谓完数是指一个整数的值等于它的因子之和。

例如，6 的因子是 1、2、3，而 6=1+2+3，故 6 是一个完数。

```
#include <stdio.h>
int main()
{
  int i,j,m,s,k,a[100] ;
  for(i=1 ; i<=1000 ; i++ )
  {
    m=i ; s=0 ; k=0 ;
    for(j=1 ; j<m ; j++)
    /***********SPACE***********/
       if(【?】)
       {
         s=s+j ;
         /***********SPACE***********/
         【?】=j ;
       }
    if(s!=0&&s==m)
```

```
    {
    /***********SPACE***********/
     for(j=0 ; 【?】 ; j++)
       printf("%4d",a[j]) ;
     printf(" =%4d\n",i) ;
    }
  }
  return 0;
}
```

（8）给定程序中，程序的功能是：从键盘输入的字符中统计数字字符的个数，用换行符结束循环。请填空。

例如：

输入：CADX2012JSJ0623

输出：8

```
#include<stdio.h>
int  main()
{
  int n=0,c;
  c=getchar();
/***********SPACE***********/
  while(【?】)
  {
/***********SPACE***********/
   if(【?】)
   n++;
   c=getchar();
  }
  printf("%d",n);
  return 0;
}
```

2. 改错性实验

（1）功能：求如下表达式：

$$s=1+\frac{1}{1+2}+\frac{1}{1+2+3}+\cdots\cdots+\frac{1}{1+2+3\cdots+n}$$

```
#include <stdio.h>
int  main()
{
    int n;
 int i,j,t;
    double s=0;
   printf("Please input a number:");
   while(i=1;i<=n;i++);
   {
    t=0;
    for(j=1;j<=i;j++)
    t=t+j;
    s=s+1/t;
   }
   print("%d",n) ;
   printf("%10.6f\n",s);
   return 0;
}
```

（2）功能：找出大于 m 的最小素数，并将其作为函数值返回。

```
#include <math.h>
#include <stdio.h>
int main()
{
  int i,k, m;
 scanf("%d",&n);
  for(i=m+1;;i++)
  {
   /**********FOUND**********/
    for(k=1;k<i;k++)
     /**********FOUND**********/
      if(i%k!=0) break;
       /**********FOUND**********/
        if(k<i)
         /**********FOUND**********/
          printf("%d\n", k);
  }
  return 0;
}
```

（3）功能：将一个正整数分解质因数。例如：输入 90，打印出 90=2*3*3*5。

```
#include <stdio.h>
int main()
{
  int n,i;
  printf("\nplease input a number:\n");
  scanf("%d",&n);
  printf("%d=",n);
  for(i=2;i<=n;i++)
  {
   /**********FOUND**********/
    while(n==i)
    {
     /**********FOUND**********/
      if(n%i==1)
      {
        printf("%d*",i);
       /**********FOUND**********/
        n=n%i;
      }
      else
        break;
    }
  }
  printf("%d",n);
  return 0;
}
```

（4）下列程序功能：输出 Fabonacci 数列的前 20 项，要求变量类型定义成浮点型，输出时只输出整数部分，输出前 20 项数。

```
#include <stdio.h>
int main()
{
  int i;
```

```
  float f1=1,f2=1,f3;
  /**********FOUND**********/
  printf("%8d",f1);
  /**********FOUND**********/
  for(i=1;i<=20;i++)
  {
    f3=f1+f2;
    /**********FOUND**********/
    f2=f1;
    /**********FOUND**********/
    f3=f2;
    printf("%8.0f",f1);
  }
  printf("\n");
  return 0;
}
```

（5）一个偶数总能表示为两个素数之和。

```
#include <stdio.h>
#include <math.h>
int main()
{
  int a,b,c,d;
  /**********FOUND**********/
  scanf("%d",a);
  for(b=3;b<=a/2;b+=2)
  {
    for(c=2;c<=sqrt(b);c++)
      if(b%c==0)
        break;
    if(c>sqrt(b))
      /**********FOUND**********/
      d=a+b;
    else
      break;
    for(c=2;c<=sqrt(d);c++)
      /**********FOUND**********/
      if(d%c=0)
    break;
    if(c>sqrt(d))
    printf("%d=%d+%d\n",a,b,d);
  }
  return 0;
}
```

（6）一个整数，它加上100之后是一个完全平方数，再加上168又是一个完全平方数，请问该整数是多少？

```
#include <stdio.h>
#include <math.h>
int main()
{
  long int i;
  double x=0.0,y;
```

```
/**********FOUND**********/
   for (i==1;i<100000;i++)
   {
     /**********FOUND**********/
     x=sqrt(i+100)
     y=sqrt(i+268);
     /**********FOUND**********/
     if(x*x==i+100||y*y==i+268)
        printf("\n%ld\n",i);
   }
   return 0;
}
```

（7）以下程序能求出 1×1+2×2+…+n×n≤1000 中满足条件的最大的 n。

```
#include <stdio.h>
#include "string.h"
int main()
{
   int n,s;
   /**********FOUND**********/
   s==n=0;
   /**********FOUND**********/
   while(s>1000)
   {
     ++n;
     s+=n*n;
   }
   /**********FOUND**********/
   printf("n=%d\n",&n-1);
   return 0;
}
```

3. 设计性实验

（1）功能：有 1、2、3、4 这四个数字，能组成多少个互不相同且无重复数字的三位数？都是多少？

（2）计算两个数的最大公约数和最小公倍数。分析：首先随机输入两个数 m、n（默认 $m>n$），接下来使 k 为 m 除以 n 的余数。如果 m 能被 n 整除，则 k 值为 0，n 为这两个数的最大公约数；否则，使 k 代替 n，n 代替 m，重复以上过程，直到 k 值为 0。要求：先画出 N-S 图，然后编写程序。

（3）编写程序实现输入整数 n，输出如下所示由数字组成的菱形。（其中 $n=5°$ ）

```
    1
   1 2 1
  1 2 3 2 1
 1 2 3 4 3 2 1
1 2 3 4 5 4 3 2 1
 1 2 3 4 3 2 1
  1 2 3 2 1
   1 2 1
    1
```

（4）从键盘上输入任意 n 个整数，求出每个整数的各位数字的平方和。

（5）哥德巴赫猜想：任意一个大于 6 的偶数都能表示为两个素数之和。要求：将 6～20 的所有偶数表示成素数之和的形式。

三、实验注意事项

1. 对于双重循环来说，外层循环往往控制变化较慢的参数（如数据项的个数，图形的层数），而内层循环变化较快（数据项的计算、图形中每行字符的数量等）。
2. 外循环改变一次，内循环变量从初值到终值执行一遍。
3. break、continue 控制循环语句结束，一般和 if 结合。break 语句跳出本层循环，continue 语句跳出本次循环。

实验 7　函数（一）

一、实验目的

1. 掌握函数的定义、调用以及函数值返回的方法。
2. 掌握函数形参和实参的区别、对应关系。

二、实验内容

1. 验证性实验

（1）计算并输出 high 以内最大的 10 个素数之和，high 由主函数传给 fun 函数，若 high 的值为 100，则函数的值为 732。

```
#include <stdio.h>
#include <math.h>
int fun( int high )
{
  int sum = 0,n=0,j,yes;
  /***********SPACE***********/
  while ((high >= 2) && (【?】))
  {
    yes = 1;
    for (j=2; j<=high/2; j++ )
      /***********SPACE***********/
      if (【?】)
      {
        yes=0;
        break;
      }
    if (yes)
    {
      sum +=high;
      n++;
    }
    high--;
  }
  /***********SPACE***********/
  【?】;
}
```

```
int main ( )
{
   printf("%d\n", fun (100));
   return 0;
}
```

（2）fun 函数的功能是计算 $s=1+1/2!+1/3!+\cdots+1/n!$，请填写程序所缺内容。

```
#include "stdio.h"
double fun(int n)
{
   double s=0.0,fac=1.0;
   int i;
   for(i=1; i<=n; i++)

   {
   /***********SPACE***********/
     fac=fac * 【?】;

   /***********SPACE***********/
     s=【?】;
   }
  return s;
}

int main()
{
    double fun(int n);
    double s;
    int t;
    scanf("%d",&t);
    s=fun(t);
    printf("s=%f\n",s);
    return 0;
}
```

（3）计算并输出500以内最大的10个能被13或17整除的自然数之和。

```
#include <stdio.h>
/***********SPACE***********/
int fun(【?】 )
{
  int m=0,  mc=0;
 /***********SPACE***********/
  while (k >= 2 && 【?】)
  {
 /***********SPACE***********/
    if (k%13 == 0 || 【?】)
    {
      m=m+k;
      mc++;
    }
    k--;
  }
 /***********SPACE***********/
  【?】;
```

```
}
int main ( )
{
  printf("%d\n", fun (500));
  return 0;
}
```

（4）sum 函数的功能为计算 1+2+3+…+n 的累加和，请填写程序所缺内容。

```
#include"stdio.h"
int sum(int n)

{
  /***********SPACE***********/
   int i,【?】;
   for(i=1;i<=n;i++)
  /***********SPACE***********/
   【?】;
   return(sum) ;
}

int  main()
{
      int sum(int n);
      int a,b;
      scanf("%d",&a);
      b=sum(a);
      printf("%d\n",b);
      return 0;
}
```

（5）以下程序求 100～200 之内的素数。

```
#include <stdio.h>
#include "math.h"
void sushu(int m)
{
     int k;
     int i;
/***********SPACE***********/
     【?】
     for(i=2;i<=k;i++)
/***********SPACE***********/
          【?】
               if(i>=k+1) printf("%4d",m);
}
int main()
{
     int m;
     for ( m=101;m<=200;m++)
/***********SPACE***********/
          【?】;
     return 0;
}
```

（6）下面程序的功能是打印出所有的水仙花数，请填写程序所缺内容。

注：水仙花数是指一个三位数的各位数字的立方和是这个数本身。

```
#include"stdio.h"
void  f( int n)
{
  int i,j,k;
  i=n/100;
  /***********SPACE***********/
  j=【?】;
  k=n%10;
  /***********SPACE***********/
  if(【?】)
  {
    printf("%5d\n",n);
  }
}
int main()
{
  void f(int n);
  int i;
  for(i=100;i<1000;i++)
   f(i);
  return 0;
}
```

（7）求 100~999 的水仙花数。

说明：水仙花数是指一个三位数的各位数字的立方和是这个数本身。

例如：$153 = 1^3 + 5^3 + 3^3$。

```
#include <stdio.h>
int fun(int n)
{ int i,j,k,m;
  m=n;
  /***********SPACE***********/
  【?】;
  for(i=1;i<4;i++)
  {
    /***********SPACE***********/
    【?】;
    m=(m-j)/10;
    k=k+j*j*j;
  }
  if(k==n)
    /***********SPACE***********/
    【?】;
  else
    return(0);}
int main()
{
  int i;
  for(i=100;i<1000;i++)
    /***********SPACE***********/
  if(【?】==1)
    printf("%d is ok!\n" ,i);
  return 0;
}
```

（8）以下程序求 100~200 的素数。

```
#include <stdio.h>
#include "math.h"
void sushu(int m)
{
    int k;
    int i;
/**********SPACE**********/
    【?】
    for(i=2;i<=k;i++)
/**********SPACE**********/
        【?】
            if(i>=k+1) printf("%4d",m);
}
int main()
{
    int m;
    for ( m=101;m<=200;m++)
/**********SPACE**********/
        【?】;
    return 0;
}
```

2. 改错性实验

（1）计算 1!+2!+……+11!。改正下列程序编译语法错误，再用设置断点的方法调试以改正逻辑错误。

```
#include <stdio.h>
int main()
{
    int i;
    long sum;
    for(i = 1; i <= 11; i++)
        sum += fact(i);//调试时，在此设置断点
    printf("1!+2!+…+11!=%f\n ", sum);
return 0;
}
long fact(int n)
{
    int i;
    long f;
    for(i=1; i <= n; i++)
        f = f * i;
    return f;//调试时，设置断点
}
```

（2）以下是一个简单的计算器，修改程序中的错误，使其能实现计算器的功能。

```
#include <stdio.h>
int main()
{
    int choice;
    double x,y;
    printf("输入两个实数:\n");
    scanf("%lf,%lf",&x , &y );
    while(1)
    {
```

```
            printf("%10s 简单计算器 \n", " ");
            printf("%10s 1 - 加法计算 \n", " ");
            printf("%10s 2 - 减法计算 \n", " ");
            printf("%10s 3 - 乘法计算 \n", " ");
            printf("%10s 4 - 除法计算 \n", " ");
            printf("%10s 5 - 退出 \n", " ");
            printf("请选择（1-5）:");
            scanf("%d",&choice);
            if( chioce >= 1 && choice < 5 ) call(choice,x,y);
            else { printf("退出计算! \n"); break;}
        }
        return 0;
    }
    void call(int choice, double x, double y)
    {
        switch(choice)
        {
            case 1: printf("和为: %.2f\n ",add(double x,double y)); break;
            case 2: printf("差为: %.2f\n ",sub(double x,double y)); break;
            case 3: printf("积为: %.2f\n ",mul(double x,double y)); break;
            case 4: printf("商为: %.2f\n ",double x,double y);
        }
    }
    void add(double x, double y )
    {   return ( x + y ); }
    void sub(double x, double y )
    {   return ( x - y ); }
    void mul(double x, double y )
    {   return ( x * y ); }
    void div(double x, double y )
    {   return ( x / y ); }
```

将上述程序中的自定义函数分别保存在以函数名为名称的头文件中，同时用#include 命令将这些头文件保存在 main 函数所在的文件中，然后编译、连接、执行。

（3）编写函数 fun，求两个整数的最小公倍数，然后用主函数 main()调用这个函数并输出结果，两个整数由键盘输入。

```
#include <stdio.h>
int fun(int m,int n)
{
  int i;
  /**********FOUND**********/
  if (m=n)
  {
    i=m;
    m=n;
    n=i;
  }
  for(i=m;i<=m*n;i+=m)
    /**********FOUND**********/
    if(i%n==1)
      return(i);
  return 0;
```

```
}
int main()
{
  unsigned int m,n,q;
  printf("m,n=");
  scanf("%d,%d",&m,&n);
  /**********FOUND**********/
  q==fun(m,n);
  printf("p(%d,%d)=%d",m,n,q);
  return 0;
}
```

（4）判断整数 x 是否是同构数。若是同构数，函数返回 1；否则返回 0。说明：所谓同构数是指这个数出现在它的平方数的右边。

例如：输入整数 25，25 的平方数是 625，25 是 625 中右侧的数，所以 25 是同构数。

x 的值由主函数从键盘读入，要求不大于 100。

```
#include <stdio.h>
#include <stdlib.h>
int fun(int x)
{
  /**********FOUND**********/
  int k
  /**********FOUND**********/
  k=x;
  /**********FOUND**********/
  if((k%10==x)&&(k%100==x)&&(k%1000==x))
    return 1;
  else
    return 0;
}
int main()
{
  int x,y;
  printf("\nPlease enter a integer numbers:");
  scanf("%d",&x);
  if(x>100){printf("data error!\n");exit(0);}
  y=fun(x);
  if(y)
    printf("%d YES\n",x);
  else
    printf("%d NO\n",x);
  return 0;
}
```

（5）编写函数 fun 计算下列分段函数的值：

$$f(x)=\begin{cases} x\times 20 & x<0 \text{且} x\neq -3 \\ \sin(x) & 0\leqslant x<10 \text{且} x\neq 2 \text{及} x\neq 3 \\ x\times x+x-1 & \text{其他} \end{cases}$$

```
#include <math.h>
#include <stdio.h>
double fun(double x)
{
  /**********FOUND**********/
  double y
  /**********FOUND**********/
  if (x<0 || x!=-3.0)
     y=x*20;
  else if(x>=0 && x<10.0 && x!=2.0 && x!=3.0)
     y=sin(x);
  else
     y=x*x+x-1;
  /**********FOUND**********/
  return x;
}
int main()
{
  double x,f;
  printf("input x=");
  scanf("%f",&x);
  f=fun(x);
  printf("x=%f,f(x)=%f\n",x,f);
  return 0;
}
```

（6）判断 m 是否为素数，若是返回 1，否则返回 0。

```
#include <stdio.h>
/**********FOUND**********/
void fun( int n)
{
  int i,k=1;
  if(m<=1)  k=0;
  /**********FOUND**********/
  for(i=1;i<m;i++)
    /**********FOUND**********/
    if(m%i=0)  k=0;
      /**********FOUND**********/
      return m;
}
int main()
{
  int m,k=0;
  for(m=1;m<100;m++)
    if(fun(m)==1)
    {
      printf("%4d",m);k++;
      if(k%5==0) printf("\n");
    }
    return 0;
}
```

（7）求 1 到 10 的阶乘的和。

```
#include <stdio.h>
int main()
{
  int i;
```

```
  float s=0;
  float fac(int n);
/**********FOUND**********/
  for(i=1;i<10;i++)
    /**********FOUND**********/
    s=fac(i);
  printf("%f\n",s);
  return 0;
}
float fac(int n)
{
/**********FOUND**********/
  int  y=1;
  int i;
  for(i=1 ;i<=n;i++)
    y=y*i;
  /**********FOUND**********/
  return;
}
```

（8）输出 Fabonacci 数列的前 20 项，要求变量类型定义成浮点型，输出时只输出整数部分，输出项数不得多于或少于 20。

```
#include <stdio.h>
fun()
{
  int i;
  float f1=1,f2=1,f3;
  /**********FOUND**********/
  printf("%8d",f1);
  /**********FOUND**********/
  for(i=1;i<=20;i++)
  {
    f3=f1+f2;
    /**********FOUND**********/
    f2=f1;
    /**********FOUND**********/
    f3=f2;
    printf("%8.0f",f1);
  }
  printf("\n");
}

int main()
{
  fun();
  return 0;
}
```

（9）计算并输出 k 以内最大的 10 个能被 13 或 17 整除的自然数之和。k 的值由主函数传入。
例如：若 k 的值为 500，则函数值为 4622。

```
#include <stdio.h>
int fun(int k)
{
  int m=0,mc=0;
  /**********FOUND**********/
  while ((k>=2)||(mc<10))
```

```
    {
      /**********FOUND**********/
      if((k%13=0)||(k%17=0))
      {
        m=m+k;
        mc++;
      }
      /**********FOUND**********/
      k++;
    }
    /**********FOUND**********/
    return ;
  }

  int main()
  {
    printf("%d\n",fun(500));
    return 0;
  }
```

（10）求出以下分数序列的前 n 项之和。和值通过函数值返回 main 函数。

$$2/1+3/2+5/3+8/5+13/8+21/13\cdots$$

例如：若 $n=5$，则应输出：8.391667。

```
#include <stdio.h>
/**********FOUND**********/
fun ( int n )
{
  int a, b, c, k; double s;
  s = 0.0; a = 2; b = 1;
  for ( k = 1; k <= n; k++ )
  {
    /**********FOUND**********/
    s = (double)a / b;
    c = a;
    a = a + b;
    b = c;
  }
  /**********FOUND**********/
  return c;
}

int main( )
{
  int n = 5;
  printf( "\nThe value of function is: %lf\n", fun ( n ) );
  return 0;
}
```

（11）编写函数 fun 求 20 以内所有 5 的倍数之积。

```
#include <stdio.h>
#define N 20
int fun(int m)
{
  /**********FOUND**********/
  int s=0,i;
  for(i=1;i<N;i++)
```

```
       /*********FOUND*********/
    if(i%m=0)
      /*********FOUND*********/
      s=*i;
  return s;
}
int main()
{
  int sum;
  sum=fun(5);
  printf("%d 以内所有%d 的倍数之积为： %d\n",N,5,sum);
  return 0;
}
```

（12）编写函数求 2!+4!+6!+8!+10!+12!+14!。

```
#include <stdio.h>
long  sum(int n)
{
  /*********FOUND*********/
  int i,j
  long    t,s=0;
  /*********FOUND*********/
  for(i=2;i<=n;i++)
  {
    t=1;
    for(j=1;j<=i;j++)
    t=t*j;
    s=s+t;
  }
  /*********FOUND*********/
  return(t);
}
int main()
{
  printf("this sum=%ld\n",sum(14));
  return 0;
}
```

（13）编写函数 fun，求 1000 以内所有 8 的倍数之和。

```
#include <stdio.h>
#define N 1000
int fun(int m)
{
  /*********FOUND*********/
  int s=0;i;
  /*********FOUND*********/
  for(i=1;i>N;i++)
    /*********FOUND*********/
    if(i/m==0)
      s+=i;
  return s;
}
int main()
{
  int sum;
  sum=fun(8);
  printf("%d 以内所有%d 的倍数之和为：%d\n",N,8,sum);
```

        return 0;
}

3. 设计性实验

（1）公式 e=1+1/1!+1/2!+1/3!+…，求 e 的近似值，精度为 10 的 -6 次方。

```
#include<stdio.h>
//函数功能：计算e,精度为f;
double fun(double f)
{
        double e=1;
        double jc=1;//求阶乘，并存入jc中
        /**********Program**********/



        /********** End **********/
        return e;
}
int main()
{
printf("e=%f\n",fun(10e-6));
 return 0;
 }
```

（2）用 do-while 语句求 1~100 的累计和。

```
#include<stdio.h>
long int  fun(int n)
{
  /**********Program**********/




  /********** End **********/
}
int main ()
{
   int i=100;
   printf("1~100的累加和为：%ld\n",fun(i));
   return 0;
}
```

（3）求一个四位数的各位数字的立方和。

```
#include <stdio.h>
int fun(int n)
{
  /**********Program**********/
```

```
/********** End **********/
}
int main()
{
  int k;
  k=fun(1234);
  printf("k=%d\n",k);
  return 0;
}
```

（4）计算并输出给定整数 n 的所有因子之和（不包括 1 与自身）。

 n 的值不大于 1000。例如：n 的值为 855 时，应输出 704。

```
#include <stdio.h>
int fun(int n)
{
  /**********Program**********/




  /********** End **********/
}
int main()
{
  printf("s=%d\n",fun(855));
  return 0;
}
```

（5）判断 m 是否为素数。

```
#include "stdio.h"
int fun(int m)
{
  /**********Program**********/



  /********** End **********/
}
int main()
{
  int m,k=0;
  for(m=100;m<200;m++)
    if(fun(m))
```

```
        {
          printf("%4d",m);
          k++;
          if(k%5==0)
            printf("\n");
        }
        printf("k=%d\n",k);
    return 0;
    }
```

（6）编写函数 fun，其功能是，根据整型形参 m，计算如下公式的值：y=1/2!+1/4!+…+1/m!（m 是偶数）。

```
#include <stdio.h>
double fun(int m)
{
  /**********Program**********/




  /********** End **********/
}
int main()
{
  int n;
  printf("Enter n: ");
  scanf("%d", &n);
  printf("\nThe result is %1f\n", fun(n));
  return 0;
}
```

三、实验注意事项

1. 定义函数时，函数名后的圆括号后面不能加";"，否则将破坏函数的整体性。
2. 在函数体内，不能再对形参进行定义和说明。
3. 实参和形参的个数与类型相同。

实验 8　函数（二）

一、实验目的

1. 掌握函数的嵌套调用和递归调用的方法。
2. 灵活运用函数的嵌套调用和递归调用。
3. 掌握局部变量与全局变量。

二、实验内容

1. 验证性实验

（1）单步调试程序，在运行过程中监测变量 x, y, a, b, max 值的变化，结合程序运行结果，说明程序的功能。

```c
int function (int x,int y)
{
    int z;
    z = x > y?x :y;
    return z;
}
 int main()
{
    int a,b,max;
    printf("\n input 2 integer numbers a,b: ");
    scanf("%d,%d ",&a, &b);
    max=function(a,b);
    printf("\n Max value is %d ",max);
    return 0;
}
```

（2）分别利用单步调试和设置断点的方法调试程序，在 watch 窗口中设置变量 i、f、result，并观察其值的变化，理解程序的功能。

```c
#include <stdio.h>
int result;   //全局变量
int fact(int n)
{
    static int f = 1;//静态局部变量
    f = f * n;
    result += f ; //调试时，可在此设置断点
    return f;
    }
int main()
{
    auto int i;//自动变量
    for(i = 1; i <= 5; i++)   //调试时，可在此设置断点
    printf("%d! = %d\n ",i, fact(i));//调试时，在此设置断点
    printf("1!+2!+…+5!=%d\n ",result);
return 0;
}
```

（3）用递归法将一个整数 n 转换成字符串，例如输入 483，应输出对应的字符串 "483"。n 的位数不确定，可以是任意位数的整数。

```c
#include <stdio.h>
void convert(int n)
{
  int i;
  /***********SPACE***********/
  if((【?】)!=0)
    convert(i);
  /***********SPACE***********/
  putchar(n%10+【?】);
}

int main()
{
```

```
  int number;
  printf("\ninput an integer:");
  scanf("%d",&number);
  printf("Output:");
  if(number<0)
  {
    putchar('-');
    /***********SPACE***********/
    【?】;
  }
  convert(number);
  return 0;
}
```

2. 改错性实验

（1）利用递归方法求 5!。

```
#include <stdio.h>
int main()
{
  int fact();
  printf("5!=%d\n",fact(5));
  return 0;
}
int fact(j)
int j;
{
  int sum;
  /**********FOUND**********/
  if(j=0)
    /**********FOUND**********/
    sum=0;
  else
    sum=j*fact(j-1);
  /**********FOUND**********/
  return j;
}
```

（2）利用递归函数调用方式，将所输入的 5 个字符，以相反顺序打印出来。

```
#include<stdio.h>
int main()
{
  int i=5;
  void palin(int n);
  printf("\40:");
  palin(i);
  printf("\n");
  return 0;
}

void palin(n)
int n;
{
  /**********FOUND**********/
  int next;
  if(n<=1)
  {
    /**********FOUND**********/
```

```
        next!=getchar();
        printf("\n\0:");
        putchar(next);
    }
    else
    {
        next=getchar();
        /**********FOUND**********/
        palin(n);
        putchar(next);
    }
}
```

（3）函数 fun 的功能是按以下递归公式求函数值。

例如：当给 n 输入 5 时，函数值为 240；当给 n 输入 3 时，函数值为 60。请改正程序中的错误，使它能得到正确结果。

 不要改动 main 函数，不得增行或删行，也不得更改程序的结构。

```
#include <stdio.h>
/***********FOUND***********/
fun(double n)
{
    int c;
/***********FOUND***********/
    if(n=1)
        c=15;
    else
        c=fun(n-1)*2;
    return(c);
}
int main()
{
    int n;
    printf("Enter n:");
    scanf("%d",&n);
    printf("The result :%d\n\n",fun(n));
    return 0;
}
```

3. 设计性实验

（1）使用递归函数求 s = 1! + 2! + 3! + … +10!。

（2）编写递归函数，计算第 n 个 Fibonacci 数。在 main 中调用该函数产生前 50 个 Fibonacci 数并输出。Fibonacci 数列中的任意第三个数是前两个数的和，即 $f(n)=f(n-1)+f(n-2)$，设 Fibonacci 数列前两个数是 1 和 1。

（3）判断一个整数 w 的各位数字平方之和能否被 5 整除，可以被 5 整除则返回 1，否则返回 0。

```
#include <stdio.h>
int fun(int w)
{
    /**********Program**********/
```

```
    /********** End **********/
}
int main()
{
  int m;
  printf("Enter m: ");
  scanf("%d", &m);
  printf("\nThe result is %d\n", fun(m));
  return 0;
}
```

（4）根据整型形参 m，计算如下公式的值：$y=1/2 + 1/4 + 1/6 + \cdots + 1/(2m)$。

例如：若 $m=9$，则应输出：1.414484

```
#include <stdio.h>
double fun(int m)
{
  /**********Program**********/




    /********** End **********/
}
int main()
{
  int n;
  printf("Enter n: ");
  scanf("%d", &n);
  printf("\nThe result is %1f\n", fun(n));
  return 0;
}
```

（5）用函数求 Fibonacci 数列前 n 项的和。

说明：Fibonacci 数列的第一项值为 1，第二项值也为 1，从第三项开始，每一项均为其前面相邻两项的和。

例如：当 $n=28$ 时，运行结果为 832039。

```
#include <stdio.h>
long sum(long f1,long f2)
{
  /**********Program**********/




    /********** End **********/
}
int main()
{
  long int f1=1,f2=1;
```

```
    printf("sum=%ld\n",sum(f1,f2));
    return 0;
}
```

（6）编写函数 fun，求任一整数 m 的 n 次方。

```
#include <stdio.h>
long fun(int m,int n)
{
  /**********Program**********/




  /********** End **********/
}
int main()
{
  int m,n;
  long s;
  long fun(int,int);
  printf("输入m和n的值:");
  scanf("%d,%d",&m,&n);
  s=fun(m,n);
  printf("s=%ld\n",s);
  return 0;
}
```

（7）编写函数 fun，求 1000 以内所有 7 的倍数之和。

```
#define N 1000
#include <stdio.h>
int main()
{
  int sum;
  sum=fun(7);
  printf("%d以内所有%d的倍数之和为：%d\n",N,7,sum);
  return 0;
}

int fun(int m)
{
  /**********Program**********/




  /********** End **********/
}
```

三、实验注意事项

1. 在函数的嵌套调用中，注意理解清楚程序的执行流程，可使用分步追踪的方法查看程序的执行过程，加深对函数调用的理解。

2. 注意全局变量、局部变量、静态局部变量重名时,在不同空间内的不同取值。

实验 9 数组(一)

一、实验目的

1. 掌握一维数组的定义、赋值、输入/输出以及引用数组元素的方法。
2. 掌握二维数组的定义、赋值、输入/输出以及引用数组元素的方法。

二、实验内容

1. 验证性实验

(1)程序填空。现有一个已按升序排好的数组,输入任意一个整数并将它插入数组中,要求还按升序排列。

```
#include <stdio.h>
int main()
{
    static int a[10]={1,7,8,17,23,24,59,62,101};
    int i,j,k,x;
    printf("输入待插入的整数:");
    scanf("%d",&x );
    _____;
    for(i=0;i< 9;i++)
if(x < a[i]) { k = i; break ; }
    for(i = 9; i>k ; i--) a[i] = a[i-1];
    _____;
    for(i=0; i<10; i++) printf("%5d",a[i]);
    printf("\n");
    return 0;
}
```

(2)程序填空。以下程序是用选择法对 10 个整数按升序排序。

```
#include <stdio.h>
/***********SPACE***********/
【?】
int main()
{
  int i,j,k,t,a[N];
  for(i=0;i<=N-1;i++)
  scanf("%d",&a[i]);
  for(i=0;i<N-1;i++)
  {
    /***********SPACE***********/
    【?】;
    /***********SPACE***********/
    for(j=i+1; 【?】;j++)
      if(a[j]<a[k]) k=j;
    /***********SPACE***********/
    if(【?】)
    {
```

```
        t=a[i];
        a[i]=a[k];
        a[k]=t;
    }
}
printf("output the sorted array:\n");
for(i=0;i<=N-1;i++)
    printf("%5d",a[i]);
printf("\n");
return 0;
}
```

（3）程序填空。产生 10 个[30，90]区间上的随机整数，然后对其用选择法进行由小到大的排序。

```
#include <stdio.h>
int main()
{
    /***********SPACE***********/
    【?】;
    int i,j,k;
    int a[10];
    for(i=0;i<10;i++)
        a[i]=random(61)+30;
    for(i=0;i<9;i++)
    {
        /***********SPACE***********/
        【?】;
        for(j=i+1;j<10;j++)
            /***********SPACE***********/
            if(【?】) k=j;
        if(k!=i)
        {
            t=a[k];
            a[k]=a[i];
            a[i]=t;
        }
    }
    /***********SPACE***********/
    for(【?】 )
        printf("%5d",a[i]);
    printf("\n");
    return 0;
}
```

（4）程序填空。将一个数组中的元素按逆序存放。

```
#include <stdio.h>
#define N 7
int main()
{
    static int a[N]={12,9,16,5,7,2,1},k,s;
    printf("\n the origanal array:\n");
    for (k=0;k<N;k++)
        printf("%4d",a[k]);
    /***********SPACE***********/
    for (k=0;k<N/2;【?】 )
    {
```

```
    s=a[k];
/***********SPACE***********/
    【?】;
/***********SPACE***********/
    【?】;
   }
   printf("\n the changed array:\n");
   for (k=0;k<N;k++)
     /***********SPACE***********/
     【?】 ("%4d",a[k]);
   return 0;
}
```

（5）功能：打印以下图形。

```
*****
 *****
  *****
   *****
    *****
#include <stdio.h>
int main ( )
{
  char a[5][9]={"        "};
  int i,j;
  for (i=0;i<5;i++)
/***********SPACE***********/
  for(j=i; 【?】;j++)
    a[i][j]='*';
/***********SPACE***********/
  for(【?】;i<5;i++)
  {
    for(j=0;j<9;j++)
    /***********SPACE***********/
    printf("%c", 【?】 );
    /***********SPACE***********/
    【?】;
   }
  return 0;
}
```

（6）求一个二维数组中每行的最大值和每行的和。

```
#include <stdio.h>
int main()
{
  int a[5][5],b[5],c[5],i,j,k,s=0;
  for(i=0;i<5;i++)
    for(j=0;j<5;j++)
      a[i][j]=random(40)+20;
  for(i=0;i<5;i++)
  {
    /***********SPACE***********/
    k=a[i][0]; 【?】 ;
    for(j=0;j<5;j++)
    {
      /***********SPACE***********/
```

```
        if(k<a[i][j]) 【?】 ;
        s=s+a[i][j];
      }
      b[i]=k;
      /***********SPACE***********/
      【?】 ;
    }
    for(i=0;i<5;i++)
    {
      for(j=0;j<5;j++)
        /***********SPACE***********/
        printf("%5d", 【?】 );
      printf("%5d%5d",b[i],c[i]);
      printf("\n");
    }
    return 0;
  }
```

（7）求出二维数组中的最大元素值。

```
  #include <stdio.h>
  max_value(m,n,array)
  /***********SPACE***********/
  int m,n,【?】;
  {
    int i,j,max;
    max=array[0][0];
    for(i=0;i<m;i++)
      for(j=0;j<n;j++)
        /***********SPACE***********/
        if(max<array[i][j]) 【?】;
    return(max);
  }
  int main()
  {
    int a[3][4]={{1,3,5,7},{2,4,6,8},{15,17,34,12}};
    /***********SPACE***********/
    printf("max value is %d\n",【?】);
    return 0;
  }
```

（8）输出 Fibonacci 数列的前 15 项，要求每行输出 5 项。Fibonacci 数列：1，1，2，3，5，8，13…。

```
  #include <stdio.h>
  int main()
  {
    /***********SPACE***********/
    int 【?】[14],i;
    fib[0]=1;fib[1]=1;
    for (i=2;i<15;i++)
      /***********SPACE***********/
      fib[i]=【?】;
    for(i=0;i<15;i++)
    {
      printf("%d\t",fib[i]);
      /***********SPACE***********/
      if ( 【?】 ) printf("\n");
    }
```

```
  return 0;
}
```

2. 改错性实验

（1）用移位法将数组 *a* 中的最后一个数移到最前面，其余数依次往后移动一个位置。

```
# include "stdio.h"
int main()
 { int i,t,a[10]={0,1,2,3,4,5,6,7,8,9};
   t=a[9];
   for(i=1;i<10;i++)
    a[i]=a[i-1];
  a[0]=t;
 printf("\n ");
 for(i=0;i<10;i++)  printf("%d ",a[i]);
 return 0;
}
```

请按以下步骤进行实验并思考。

① 分析程序及其特性。

② 上机运行程序，查看运行结果是否正确。

③ 用单步跟踪查找错误原因。

④ 改正错误后重新运行程序，直到结果正确为止。

⑤ 如果要用三次循环移位来实现将最后三个数移到前面，其余数依次往后移三个位置，则程序应该如何修改。

（2）输入 *n* 个学生的单科成绩，然后按从高到低的顺序排序输出。

以下是用选择法实现的排序。

```
# include "stdio.h"
int main( )
{
    int i,j,t,n,a[n];
    printf("\n n=? ");
    scanf("%d ",&n);
    printf("input n numbers :\n");
    for (i=0;i<n;i++) scanf("%d",&a[i]);
    for(i=0;i<n-1;i++)
        for(j=i+1;j<n;j++)
            if(a[i]<a[j])
            {t=a[i];a[i]=a[j];a[j]=t;}
    printf("the sorted numbers:\n");
    for(i=0;i<n;i++)  printf("%4d",a[i]);
    return 0;
}
```

请按照以下步骤进行实验并思考。

① 分析程序及其特性。

② 上机编译程序（数组 *a* 的长度可比 *n* 大些），程序是否有语法错误，应如何修改。改正错误后重新编译和运行程序，直到结果正确为止。

③ 你对选择排序算法的实现过程是否清楚，若不清楚,请用动态跟踪的方法观察其实现过程。

④ 输入冒泡排序程序，用动态跟踪观察其实现过程。

⑤ 如果要用三次循环移位来实现将最后三个数移到前面，其余数依次往后移三个位置，则程

序应该如何修改？

（3）计算数组元素中值为正数的平均值（不包括0）。

例如：数组中元素的值依次为39，-47，21，2，-8，15，0，
则程序的运行结果为19.250000。

```c
#include <stdio.h>
double fun(int s[])
{
  /**********FOUND**********/
  int  sum=0.0;
  int c=0,i=0;
  /**********FOUND**********/
  while(s[i] =0)
  {
    if (s[i]>0)
    {
      sum+=s[i];
      c++;
    }
    i++;
  }
  /**********FOUND**********/
  sum\=c;
  /**********FOUND**********/
  return c;
}
int main()
{
  int x[1000];int i=0;
  do
  {
    scanf("%d",&x[i]);}
    while(x[i++]!=0);
    printf("%f\n",fun(x));
   return 0;
}
```

（4）先从键盘上输入一个3行3列矩阵的各个元素的值，然后输出主对角线上的元素之和sum。

```c
#include <stdio.h>
void fun()
{
  int a[3][3],sum;
  int i,j;
  /**********FOUND**********/
  a=0;
  for(i=0;i<3;i++)
    for(j=0;j<3;j++)
      /**********FOUND**********/
      scanf("%d",a[i][j]);
  for(i=0;i<3;i++)
    /**********FOUND**********/
    sum=sum+a[i][j];
  /**********FOUND**********/
  printf("sum=%f\n",sum);
}

int main()
```

```
    {
        fun();
        return 0;
    }
```

（5）从键盘输入十个学生的成绩，统计最高分、最低分和平均分。max 代表最高分，min 代表最低分，avg 代表平均分。

```
#include <stdio.h>
int main( )
{
  int i;
  /***********FOUND***********/
  float a[8],min,max,avg;
  printf("input 10 score:");
  for(i=0;i<=9;i++)
  {
    printf("input a score of student:");
    /***********FOUND***********/
    scanf("%f",a);
  }
  /***********FOUND***********/
  max=min=avg=a[1];
  for(i=1;i<=9;i++)
  {
    /***********FOUND***********/
    if(min<a[i])
      min=a[i];
    if(max<a[i])
      max=a[i];
    avg=avg+a[i];
  }
  avg=avg/10;
  printf("max:%f\nmin:%f\navg:%f\n",max,min,avg);
  return 0;
}
```

（6）有一数组内放 10 个整数，要求找出最小数和它的下标，然后把它和数组中最前面的元素即第一个数对换位置。

```
#include <stdio.h>
int main( )
{
  int  i,a[10],min,k=0;
  printf("\n please input array 10 elements\n");
  for(i=0;i<10;i++)
    /***********FOUND***********/
    scanf("%d", a[i]);
  for(i=0;i<10;i++)
    printf("%d",a[i]);
  min=a[0];
  /***********FOUND***********/
  for(i=3;i<10;i++)
    /***********FOUND***********/
    if(a[i]>min)
    {
      min=a[i];
      k=i;
```

```
    }
    /**********FOUND**********/
    a[k]=a[i];
    a[0]=min;
    printf("\n after eschange:\n");
    for(i=0;i<10;i++)
      printf("%d",a[i]);
    printf("\nk=%d\nmin=%d\n",k,min);
    return 0;
}
```

（7）输入 10 个数，要求输出这 10 个数的平均值。

```
#include <stdio.h>
float average(float array[10])
{
  int i;
  float aver,sum=array[0];
  /**********FOUND**********/
  for(i=0;i<10;i++)
    sum=sum+array[i];
  aver=sum/10.0;
  return(aver);
}
int main( )
{
  /**********FOUND**********/
  int score[10],aver ;
  int i;
  printf("input 10 scores:\n");
  for(i=0;i<10;i++)
    /**********FOUND**********/
    scanf("%f", score);
  printf("\n");
  /**********FOUND**********/
  aver=average(score[10]);
  printf("average score is %5.2f",aver);
return 0;
}
```

（8）请编写函数 fun，对长度为 8 个字符的字符串，将 8 个字符按降序排列。

例如：原来的字符串为 CEAedcab，排序后输出为 edcbaECA。

```
#include<stdio.h>
#include<ctype.h>
void fun(char *s,int num)
{
  /**********FOUND**********/
  int i;j;
  char t;
  for(i=0;i<num;i++)
    /**********FOUND**********/
    for(j=i;j<num;j++)
      /**********FOUND**********/
      if(s[i]>s[j])
      {
        t=s[i];
        s[i]=s[j];
        s[j]=t;
      }
```

```
}
int main()
{
  char s[10];
  printf("输入8个字符的字符串:");
  gets(s);
  fun(s,8);
  printf("\n%s",s);
  return 0;
}
```

（9）求出数组中最大数和次最大数，并把最大数和 a[0] 中的数对调、次最大数和 a[1] 中的数对调。

```
#include <stdio.h>
#define N 20
void fun ( int * a, int n )
{
  int i, m, t, k ;
  for(i=0;i<2;i++)
  {
    /**********FOUND**********/
    m=0;
    /**********FOUND**********/
    for(k=1;k<n;k++)
      /**********FOUND**********/
      if(a[k]>a[m]) k=m;
       t=a[i];a[i]=a[m];a[m]=t;
  }
}

int main( )
{
  int b[N]={11,5,12,0,3,6,9,7,10,8}, n=10, i;
  for ( i=0; i<n; i++ ) printf("%d ", b[i]);
    printf("\n");
  fun ( b, n );
  for ( i=0; i<n; i++ )
    printf("%d ", b[i]);
  printf("\n");
  return 0;
}
```

3. 设计性实验

（1）输出 50 个 200～300 之间的随机整数，找出其中的所有素数，并按升序排列。

（2）求出 1000 以内前 20 个不能被 2、3、5、7 整除的数之和。要求：使用程序中定义的变量。

```
#include"stdio.h"
//fun 函数功能：求求出出 1000 以内，前 n 个不能被 2,3,5,7 整除的数，求出这些数的和
int fun(int n)
{
  int i,j=0,a[20],sum=0; //前 20 个不能被 2,3,5,7 整除的数保存在 a 数组中，它们的和保存在 sum 中。
       /**********Program**********/




       /********** End **********/
  return sum;
}
```

```
int main()
{
    printf("和为: %d\n",fun(20));
    return 0;
}
```

（3）C 语言编程题实现以下功能。

① 从键盘输入 10 个学生的成绩，计算平均成绩 ave。

② 统计及格人数 pass，计算高于平均分的学生的人数 better。

③ 将 10 个学生成绩排名。

```
#include "stdio.h"
#define N 10
int main()
{
    int i,j,a[N],ave,sum=0,pass=0,better=0,t;
    /**********Program**********/



    /********** End **********/
    printf("平均分:%d\n及格人数%d\n高于平均分人数%d\n",ave,pass,better);
    return 0;
}
```

（4）求出 $N \times M$ 整型数组的最大元素及其所在的行坐标及列坐标（如果最大元素不唯一，选择位置在最前面的一个）。

例如：输入的数组为

```
          1    2    3
          4   15    6
         12   18    9
         10   11    2
```

求出的最大数为 18，行坐标为 3，列坐标为 2。

```
#define N 4
#define M 3
#include <stdio.h>
int Row,Col;
int fun(int array[N][M])
{
  /**********Program**********/




  /********** End **********/
}

int main()
```

```
{
  int a[N][M],i,j,max;
  printf("input a array:");
  for(i=0;i<N;i++)
    for(j=0;j<M;j++)
      scanf("%d",&a[i][j]);
  for(i=0;i<N;i++)
  {
    for(j=0;j<M;j++)
      printf("%d",a[i][j]);
    printf("\n");
  }
  max=fun(a);
  printf("max=%d,row=%d,col=%d",max,Row,Col);
  return 0;
}
```

（5）求一批数中最大值和最小值的差。

```
#define N 30
#include <stdio.h>
int max_min(int a[],int n)
{
  /**********Program**********/




  /********** End **********/
}
int main()
{
  int a[N],i,k;
  for(i=0;i<N;i++)
    a[i]=random(51)+10;
  for(i=0;i<N;i++)
  {
    printf("%5d",a[i]);
    if((i+1)%5==0) printf("\n");
  }
  k=max_min(a,N);
  printf("the result is:%d\n",k);
  return 0;
}
```

（6）编写函数fun，对主程序中用户输入的具有10个数据的数组 *a* 按由大到小排序，并在主程序中输出排序结果。

```
#include <stdio.h>
int fun(int array[], int n)
{
  /**********Program**********/
```

```
  /********** End **********/
}

int main()
{
  int a[10],i;
  printf("请输入数组 a 中的十个数:\n");
  for (i=0;i<10;i++)
    scanf("%d",&a[i]);
  fun(a,10);
  printf("由大到小的排序结果是:\n");
  for (i=0;i<10;i++)
    printf("%4d",a[i]);
  printf("\n");
  return 0;
}
```

（7）编写函数 fun，将一个数组中的值按逆序存放，并在 main()函数中输出。

例如：原来存顺序为 8, 6, 5, 4, 1。要求改为：1, 4, 5, 6, 8。

```
#include <stdio.h>
#define N 5
void fun(int arr[],int n)
{
  /**********Program**********/




  /********** End **********/
}
int main()
{
  int a[N]={8,6,5,4,1},i;
  for(i=0;i<N;i++)
    printf("%4d",a[i]);
  printf("\n");
  fun(a,N);
  for(i=0;i<N;i++)
    printf("%4d",a[i]);
  return 0;
}
```

（8）产生 20 个[30,120]上的随机整数放入二维数组 a[5][4]中，求每行元素的和。

```
#include "stdlib.h"
#include <stdio.h>
void row_sum(int a[5][4],int b[5])
{
  /**********Program**********/
```

```
       /********** End **********/
      }
    int main()
    {
      void row_sum();
      int a[5][4],b[5],i,j;
      for(i=0;i<5;i++)
        for(j=0;j<4;j++)
          a[i][j]=random(120-30+1)+30;
        for(i=0;i<5;i++)
        {
          for(j=0;j<4;j++)
            printf("%5d",a[i][j]);
          printf("\n");
        }
        row_sum(a,b);
        for(i=0;i<5;i++)
          printf("%6d",b[i]);
        printf("\n");
      return 0;
     }
```

三、实验注意事项

1. C语言规定，数组元素的下标下界为0，因此数组元素的下标上界是该数组元素的个数减1。
2. 数值型数组要对多个数组元素赋值时，使用循环语句，使数组元素的下标依次变化，从而为每个数组元素赋值。

```
int a[10],i;
for (i=0;i<10;i++)
    scanf("%d",&a[i]);
```

3. 要对二维数组的多个数组元素赋值，应使用循环语句，并在循环过程中使数组元素的下标变化。可使用下面的方法为所有数组元素赋值：

```
int a[5][4],b[5],i,j;
  for(i=0;i<5;i++)
    for(j=0;j<4;j++)
scanf("%d",&a[i][j]);
```

实验 10　数组（二）

一、实验目的

1. 掌握字符数组和字符串函数的使用。
2. 掌握与数组有关的算法（特别是排序算法）。

二、实验内容

1. 验证性实验

（1）删除字符串中的指定字符，字符串和要删除的字符均由键盘输入。

```
#include <stdio.h>
int main()
{
  char str[80],ch;
  int i,k=0;
  /**********SPACE**********/
  gets(【?】);
  ch=getchar();
  /**********SPACE**********/
  for(i=0;【?】;i++)
    if(str[i]!=ch)
    {
       /**********SPACE**********/
       【?】;
       k++;
    }
  /**********SPACE**********/
  【?】;
  puts(str);
  return 0;
}
```

（2）将两个字符串连接为一个字符串，不许使用库函数 strcat。

```
#include <stdio.h>
#include "string.h"
JOIN(s1,s2)
char s1[80],s2[40];
{
  int i,j;
  /**********SPACE**********/
  【?】;
  /**********SPACE**********/
  for (i=0; 【?】'\0';i++)
    s1[i+j]=s2[i];
  /**********SPACE**********/
  s1[i+j]= 【?】 ;
}
int main ( )
{
  char str1[80],str2[40];
  gets(str1);gets(str2);
  puts(str1);puts(str2);
  /**********SPACE**********/
  【?】;
  puts(str1);
  return 0;
}
```

（3）删除一个字符串中的所有数字字符。

```
#include <stdio.h>
void delnum(char *s)
{
  int i,j;
  /***********SPACE***********/
  for(i=0,j=0; 【?】'\0' ;i++)
  /***********SPACE***********/
  if(s[i]<'0'【?】 s[i]>'9')
  {
    /***********SPACE***********/
    【?】;
    j++;
  }
  s[j]='\0';
}
int main ()
{
  char *item;
  printf("\n input a string:\n");
  item="";
  gets(item);
  /***********SPACE***********/
  【?】;
  printf("\n%s",item);
  return 0;
}
```

（4）统计一个字符串中的字母、数字、空格和其他字符的个数。

```
#include <stdio.h>
void fun(char s[],int b[])
{
  int i;
  for (i=0;s[i]!='\0';i++)
  if ('a'<=s[i]&&s[i]<='z'||'A'<=s[i]&&s[i]<='Z')
    b[0]++;
  /***********SPACE***********/
  else if (【?】)
    b[1]++;
  /***********SPACE***********/
  else if (【?】 )
    b[2]++;
  else
    b[3]++;
}

int main ()
{
  char s1[80];int a[4]={0};
  int k;
  /***********SPACE***********/
  【?】;
  gets(s1);
  /***********SPACE***********/
  【?】;
  puts(s1);
```

```
    for(k=0;k<4;k++)
      printf("%4d",a[k]);
    return 0;
}
```

（5）下面程序的功能是将一个字符串 str 的内容颠倒过来，请填写程序所缺内容。

```
#include "string.h"
#include "stdio.h"
int main( )
{
    int  i, j, k ;
    char  str[ ]= "1234567";
    for(i=0, j=strlen(str);i<j;i++,j--)

    {
    /***********SPACE***********/
         【?】;
    /***********SPACE***********/
         【?】;
         str[j-1]=k;
    }
    /***********SPACE***********/
         puts(【?】);
    return 0;
}
```

（6）下列程序从键盘输入一字符串（可以含有空格），再从键盘输入一个该字符串中的字符，删除该字符后重新输出字符串。

```
#include<stdio.h>
#include<string.h>
int main()
{char line[80] ;
 char ch;
 int i,j;
 int len;
 printf("输入一行字符 \n");
/***********SPACE***********/
  【?】;
 printf("输入要删除字符 ");
 ch=getchar();
 i=0;
/***********SPACE***********/
 while(【?】)
   {
   while( line[i]!=ch&&line[i]!='\0')
      i++;
      len=strlen(line);
        for(j=i;j<len-1;j++)
/***********SPACE***********/
        【?】;
        line[j]='\0';
   }
   puts(line);
   return 0;
}
```

（7）功能：考查字符串数组的应用。输出 26 个英文字母。

```
#include <stdio.h>
int main ()
{
  char string[256];
  int i;
  /***********SPACE***********/
  for (i = 0; i < 26; 【?】)
  /***********SPACE***********/
    string[i] = 【?】;
  string[i] = '\0';
  /***********SPACE***********/
  printf ("the arrary contains %s\n",【?】);
  return 0;
}
```

2. 改错性实验

（1）实现两个字符串的连接。

例如：输入 dfdfqe 和 12345 时，则输出 dfdfqe12345。

```
#include <stdio.h>
int main()
{
  char s1[80],s2[80];
  void scat(char s1[],char s2[]);
  gets(s1);
  gets(s2);
  scat(s1,s2);
  puts(s1);
  return 0;
}
void scat (char s1[],char s2[])
{
  int i=0,j=0;
  /**********FOUND**********/
  while(s1[i]= ='\0')
    i++;
  /**********FOUND**********/
  while(s2[j]= ='\0')
  {
    /**********FOUND**********/
    s2[j]=s1[i];
    i++;
    j++;
  }
  /**********FOUND**********/
  s2[j]='\0';
}
```

（2）将 s 所指字符串的反序和正序进行连接，形成一个新串放在 t 所指的数组中。

例如：当 s 所指的字符串的内容为 " ABCD " 时，t 所指数组中的内容为 " DCBAABCD "。

```
#include <stdio.h>
#include <string.h>
/**********FOUND**********/
void fun (char s, char t)
{
```

```
    int  i, d;
/**********FOUND**********/
    d = len(s);
/**********FOUND**********/
    for (i = 1; i<d; i++)
       t[i] = s[d - 1 - i];
    for (i = 0; i<d; i++)
       t[ d + i ] = s[i];
/**********FOUND**********/
    t[2*d] = '/0';
}
int main()
{
    char  s[100], t[100];
    printf("\nPlease enter string S:");
    scanf("%s", s);
    fun(s, t);
    printf("\nThe result is: %s\n", t);
    return 0;
}
```

（3）下列程序的功能为：输入一个字符串，并将其中的字符 'a' 用字符串'shu'替代后输出。例如输入为"123abcaHello"，则输出为"123shubcshuHello"。请修改错误。

```
#include <stdio.h>
int main()
{
    int i; char line[81];
/**********FOUND**********/
    scanf("%s", &line);
/**********FOUND**********/
    for(i=0; line[i]!='\n'; i++)
/**********FOUND**********/
        if(line[i]='a')
           printf("shu");
        else
/**********FOUND**********/
           printf("%s", line[i]);
    return 0;
}
```

3. 设计性实验

（1）从字符串 s 中删除指定的字符 c。

```
#include <stdio.h>
fun(char s[],char c)
{
  /**********Program**********/




  /********** End **********/
}
```

```
int main()
{
  static char str[]="turbo c and borland c++";
  char c='a';
  fun(str,c);
  printf("str=%s\n",str);
  return 0;
}
```

（2）找出一批正整数中的最大的偶数。

```
#include <stdio.h>
int fun(int a[],int n)
{

  /**********Program**********/




  /********** End **********/

}
int main()
{
  int a[]={1,2,9,24,35,18},k;
  k=fun(a,6);
  printf("max=%d\n",k);
  return 0;
}
```

（3）求大于 lim（lim 小于 100 的整数）并且小于 100 的所有素数并放在 aa 数组中，该函数返回所求出素数的个数。

```
#include<stdio.h>
#define MAX 100
int fun(int lim,int aa[MAX])
{
  /**********Program**********/




  /********** End **********/
}
int main()
{
  int limit,i,sum;
  int aa[MAX];
  printf("Please input ainteger:");
  scanf("%d",&limit);
  sum=fun(limit,aa);
  for(i=0;i<sum;i++)
  {
```

```
    if(i%10==0&&i!=0) printf("\n");
    printf("%5d",aa[i]);
  }
  return 0;
}
```

（4）用函数求一个 N 阶方阵右下三角元素的和（包括副对角线上的元素）。

```
#include <stdlib.h>
#include <stdio.h>
#define N 3
int sum(int a[][N])
{
  /**********Program**********/





  /********** End **********/
}

int main()
{
  int a[N][N],i,j;
  for(i=0;i<N;i++)
  {
    for(j=0;j<N;j++)
    {
      a[i][j]=random(10)+10;
      printf("%3d",a[i][j]);
    }
    printf("\n");
  }
  printf("==================\n");
  printf("sum=%5d\n",sum(a));
  return 0;
}
```

（5）用函数实现字符串的复制，不允许用 strcpy() 函数。

```
#include <stdio.h>
void copy(char str1[],char str2[])
{
  /**********Program**********/





  /********** End **********/
}

int main()
{
  void copy();
  char c1[40],c2[40];
  gets(c1);
```

```
    copy(c1,c2);
    puts(c2);
}
```

（6）任输入十个国家的名字，按英文字母由小到大排序。

```
#include <stdio.h>
#include <string.h>
int main( )
{
    char a[10][20],b[20]; //用来保存国家的名字 b 表示中间变量
    int i,j;   //循环变量

/**********Program**********/




/********** End **********/
    for (i=0;i<10;i++)
      printf("%s\n",a[i]);
    return 0;
}
```

（7）求 4×4 整型数组的主对角线元素的和，请编 fun 函数。

```
#include <stdio.h>
int fun(int a[4][4])
{

    /**********Program**********/


    /********** End **********/
}
int main()
{
    int a[4][4], s, i, j;

    for(i=0; i < 4; i ++)
       for(j = 0; j < 4;j ++)
    scanf("%d", &a[i][j]);
    s=fun(a);
    printf("%d",s);
}
```

（8）从键盘上输入一个字符串，放在数组中，要求将字符串中的大写字母转换成小写字母，小写字母转换成大写字母，非字母字符不变，并输出。

```
#include <stdio.h>
#include <string.h>
#define N 80

char fun(char a[])
{

    /**********Program**********/
```

```
        /**********  End  **********/
}
int  main()
{
    char a[80];
    char s;
    gets(a);
    s=fun(a);
    puts(a);
    return 0;
}
```

三、实验注意事项

1. C 语言中字符串是作为一维数组存放在内存中的，并且系统对字符串常量自动加上一个 '\0'作为结束符，所以在定义一个字符数组并初始化时要注意数组的长度。
2. 注意用 scanf 函数对字符数组整体赋值的形式。

实验 11　指针（一）

一、实验目的

1. 掌握指针的基本概念。
2. 掌握指针变量的使用。
3. 掌握指针与数组的关系。

二、实验内容

1. 验证性实验

（1）程序填空。求数组 a 中 10 个元素的和。

```
#include <stdio.h>
int main()
{
        int a[10]={1,2,3,4,5,6,7,8,9,10} ; *p,sum = 0;
        for ( p=a , p < a + 10 , p++ )
        sum = sum + _____;
        printf(" sum = % d \n",sum );
        return 0;
}
```

（2）程序填空。输入 3 个数 a, b, c，按从小到大的顺序输出。

```
#include <stdio.h>
int main()
{
  void swap(int *p1, int *p2);
  int n1,n2,n3;
  int *pointer1,*pointer2,*pointer3;
```

```
    printf("please input 3 number:n1,n2,n3:");
    scanf("%d,%d,%d",&n1,&n2,&n3);
    pointer1=&n1;
    pointer2=&n2;
    pointer3=&n3;
    /**********SPACE**********/
    if(【?】) swap(pointer1,pointer2);
    /**********SPACE**********/
    if(【?】) swap(pointer1,pointer3);
    /**********SPACE**********/
    if(【?】) swap(pointer2,pointer3);
    printf("the sorted numbers are:%d,%d,%d\n",n1,n2,n3);
    return 0;
}
/**********SPACE**********/
void swap(【?】)
int *p1,*p2;
{
    int p;
    p=*p1;*p1=*p2;*p2=p;
}
```

（3）程序填空。输出两个整数中大的那个数，两个整数由键盘输入。

```
#include <stdio.h>
#include <stdlib.h>
int main()
{
    int *p1,*p2;
    /**********SPACE**********/
    p1=【?】malloc(sizeof(int));
    p2=(int*)malloc(sizeof(int));
    /**********SPACE**********/
    scanf("%d%d",【?】,p2);
    if(*p2>*p1) *p1=*p2;
    free(p2);
    /**********SPACE**********/
    printf("max=%d\n",【?】);
    return 0;
}
```

（4）下面程序的功能是输出数组 s 中最大值元素的下标，请填写程序所缺内容。

```
#include"stdio.h"
int main( )
{
    int k, p;
    int s[ ]={1,9,7,2,10,3};
    /**********SPACE**********/
    for(p=0,k=p; p<6; 【?】)
    /**********SPACE**********/
    if(s[p]>s[k]) 【?】;
    printf("%d\n" ,k);
    return 0;
```

（5）求一批数据（数组）的最大值并返回下标。

```
#include <stdio.h>
int max(int *p,int n,int *index)
{
  int i,in=0,m;
/***********SPACE***********/
【?】;
  /***********SPACE***********/
  for (【?】;i<n;i++)
     if(m<*(p+i))
     {
       m=*(p+i);
/***********SPACE***********/
【?】;
     }
   *index=in;
/***********SPACE***********/
   【?】;
}
int main()
{
  int i,a[10]={3,7,5,1,2,8,6,4,10,9},m;
/***********SPACE***********/
  m=【?】;
/***********SPACE***********/
  printf("最大值%d,下标%d",【?】,i);
  return 0;
}
```

2. 改错性实验

（1）修改程序中的错误。下列程序用来统计数组 a 中所有偶数的个数。

```
#include <stdio.h>
int main()
{
        int a[10],*p, n = 0;
        printf(" 请输入 10 个整数：\n" );
        for ( p=a, p < a + 10,p++ )
            scanf("%d ",*p );
        for( ; p>=a ; p - - )
        if( p % 2 = =0 )  n+ + ;
        printf(" 偶数的个数为 % d \n",n );
        return 0;
}
```

（2）将 s 所指字符串中的字母转换为按字母序列的后续字母（但 Z 转换为 A， z 转换为 a），其他字符不变。

```
#include <stdio.h>
void fun (    char *s)
{
  /***********FOUND***********/
  while(*s!="\\0")
  {
```

```c
      if(*s>='A' && *s <= 'Z' || *s >= 'a' && *s<='z')
      {
        if(*s=='Z')
          *s='A';
        else if(*s=='z')
          *s='a';
        else
        /**********FOUND**********/
          s += 1;
      }
    /**********FOUND**********/
      s++
  }
}

int main()
{
  char s[80];
  printf("\n Enter a string with length < 80. :\n\n ");
  gets(s);
  printf("\n The string : \n\n ");
  puts(s);
  fun ( s );
  printf ("\n\n The Cords :\n\n ");
  puts(s);
  return 0;
}
```

（3）在一个一维整型数组中找出其中最大的数及其下标。

```c
#include <stdio.h>
#define N 10
/**********FOUND**********/
float fun(int *a,int *b,int n)
{
  int *c,max=*a;
  for(c=a+1;c<a+n;c++)
    if(*c>max)
    {
      max=*c;
      /**********FOUND**********/
      b=c-a;
    }
  return max;
}

int main()
{
  int a[N],i,max,p=0;
  printf("please enter 10 integers:\n");
  for(i=0;i<N;i++)
    /**********FOUND**********/
    get("%d",a[i]);
  /**********FOUND**********/
```

```
  m=fun(a,p,N);
  printf("max=%d,position=%d",max,p);
  return 0;
}
```

（4）在一个已按升序排列的数组中插入一个数，插入后，数组元素仍按升序排列。

```
#include <stdio.h>
#define N 11
int main()
{
  int i,number,a[N]={1,2,4,6,8,9,12,15,149,156};
  printf("please enter an integer to insert in the array:\n");
  /**********FOUND**********/
  scanf("%d",&number)
  printf("The original array:\n");
  for(i=0;i<N-1;i++)
    printf("%5d",a[i]);
  printf("\n");
  /**********FOUND**********/
  for(i=N-1;i>=0;i--)
    if(number<=a[i])
  /**********FOUND**********/
    a[i]=a[i-1];
  else
  {
    a[i+1]=number;
    /**********FOUND**********/
    exit;
  }
  if(number<a[0]) a[0]=number;
    printf("The result array:\n");
  for(i=0;i<N;i++)
    printf("%5d",a[i]);
  printf("\n");
  return 0;
}
```

3. 设计性实验

（1）从键盘输入10个整数存放在一维数组中。首先将数组内容倒置，然后输出。用数组的指针实现。

（2）求5行5列二维数组的转置矩阵并输出。用二维数组的指针来实现。

（3）编写函数实现两个数据的交换，在主函数中输入任意三个数据，调用函数对这三个数据从大到小排序。

```
#include<stdio.h>
void swap(int *a,int *b)
{
  /**********Program**********/
```

```
      /********** End **********/
}
int main()
{
  int x,y,z;
  scanf("%d%d%d",&x,&y,&z);
  if(x<y)swap(&x,&y);
  if(x<z)swap(&x,&z);
  if(y<z)swap(&y,&z);
  printf("%3d%3d%3d",x,y,z);
  return 0;
}
```

（4）实现两个整数的交换。

例如：给 a 和 b 分别输入 60 和 65，输出为：a=65, b=60

```
#include<stdio.h>
void fun(int *a,int *b)
{
  /**********Program**********/




      /********** End **********/
}
int main()
{
  int a,b;
  printf("Enter a,b:");
  scanf("%d%d",&a,&b);
  fun(&a,&b);
  printf("a=%d b=%d\n",a,b);
  return 0;
}
```

三、实验注意事项

1. 取地址运算符"&"和指针运算符"*"的区别。

2. 对于指向一维数组元素的指针变量 p 和数组名 a，p++代表 p 指向数组的下一个元素，而 a++是不能实现的，因为数组名 a 是一个指针型变量，它代表数组的首元素地址。

实验 12 指针（二）

一、实验目的

1. 掌握指针与字符串的关系。

2. 掌握指针数组、动态数组的用法，熟悉指向指针的指针。

二、实验内容

1. 验证性实验

（1）程序填空。将 s 所指字符串的正序和反序进行连接，形成一个新串放在 t 所指的数组中。

例如：当 s 串为 ABCD 时，则 t 串的内容应为 ABCDDCBA。

```
#include <stdio.h>
#include <string.h>
void fun (char *s, char *t)
{
  int   i, d;
  /***********SPACE***********/
  d = 【?】;
  /***********SPACE***********/
  for (i = 0; i<d; 【?】)
    t[i] = s[i];
  for (i = 0; i<d; i++)
    /***********SPACE***********/
    t[【?】] = s[d-1-i];
  /***********SPACE***********/
  t[【?】] ='\0';
}
int main()
{
  char  s[100], t[100];
  printf("\nPlease enter string S:");
scanf("%s", s);
  fun(s, t);
  printf("\nThe result is: %s\n", t);
  return 0;
}
```

（2）程序填空。功能是实现字符串的连接，即将 t 所指字符串复制到 s 所指字符串的尾部。

例如：s 所指字符串为 abcd，t 所指字符串为 efgh，函数调用后 s 所指字符串为 abcdefgh。

```
#include <stdio.h>
#include <string.h>
int   main()
{
    char   s[128]   =    "abcd";
    char   t[128]   =    "efgh";
    char*  ss   =   &s[strlen(s)];
    char*  tt   =   t;
/***********SPACE***********/
    while  (【?】)
/***********SPACE***********/
          *ss++  =   *【?】;
    *++ss  =  '\0';
    printf("%s\n",s );
    return 0;
}
```

（3）程序填空。删除字符串中的数字字符。

例如：输入字符串 48CTYP9E6，则输出 CTYPE。

```
#include <stdio.h>
/**********SPACE**********/
void fun (【?】)
{
  char *p=s;
  while(*p)
    if((*p>='0')&&(*p<='9')) p++;
    /**********SPACE**********/
    else *s++=【?】;
    /**********SPACE**********/
    【?】;
}

int main( )
{
  char item[100] ;
  printf("\nEnter a string: ");
  gets(item); fun(item);
  printf("\nThe string:\"%s\"\n",item);
  return 0;
}
```

（4）程序填空。将数组 s2 中的数字字符拼接到数组 s1 后面。

```
#include "stdio.h"
int main()
{
  char s1[20]="xy",s2[]="ab12DFc3G",*t1=s1,*t2=s2;
  while(*t1!='\0')
/**********SPACE**********/
   【?】;
  while(*t2!='\0')
  {
    if(*t2>='0'&&*t2<='9')
    {
/**********SPACE**********/
     *t1=【?】;
      t1++;
    }
    t2++;
  }
/**********SPACE**********/
  *t1=【?】;
  puts(s1);
  return 0;
}
```

2. 改错性实验

（1）输入一行英文文本，将每一个单词的第一个字母变成大写。

例如：输入 "This is a C program."，输出为 "This is A C Program."。

```
#include <string.h>
#include <stdio.h>
/**********FOUND**********/
fun(char p)
{
  int k=0;
```

```
/**********FOUND**********/
do while(*p=='\0')
{
  if(k==0&&*p!=' ')
  {
    *p=toupper(*p);
    /**********FOUND**********/
    k=0;
  }
  else if(*p!=' ')
    k=1;
  else
    k=0;
  /**********FOUND**********/
  *p+;
}
}
int main()
{
  char str[81];
  printf("please input a English text line:");
  gets(str);
  printf("The original text line is :");
  puts(str);
  fun(str);
  printf("The new text line is :");
  puts(str);
  return 0;
}
```

（2）将一个字符串中的大写字母转换成小写字母。例如：输入"aSdFG"，输出为"asdfg"。

```
#include<stdio.h>
/**********FOUND**********/
bool fun(char *c)
{
  if(*c<='Z'&&*c>='A')*c-='A'-'a';
  /**********FOUND**********/
  fun= c;
}
int main()
{
  /**********FOUND**********/
  char s[81],p=s;
  gets(s);
  while(*p)
  {
    *p=fun(p);
    /**********FOUND**********/
    puts(*p);
    p++;
  }
  putchar('\n');
  return 0;
}
```

（3）用指针作函数参数，编程序求一维数组中的最大和最小的元素值。

```
#include <stdio.h>
#define N 10
/***********FOUND***********/
void maxmin(int arr[ ],int *pt1, *pt2, n)
{
  int i;
  /***********FOUND***********/
  *pt1=*pt2=&arr[0];
  for(i=1;i<n;i++)
  {
    /***********FOUND***********/
    if(arr[i]<*pt1)  *pt1=arr[i];
    if(arr[i]<*pt2)  *pt2=arr[i];
  }
}
int main( )
{
  int array[N]={10,7,19,29,4,0,7,35,-16,21},*p1,*p2,a,b;
  /***********FOUND***********/
  *p1=&a;*p2=&b;
  maxmin(array,p1,p2,N);
  printf("max=%d,min=%d",a,b);
  return 0;
}
```

（4）删除字符串 s 中的所有空白字符（包括 Tab 字符、回车符及换行符）。输入字符串时用'#'结束输入。

```
#include <string.h>
#include <stdio.h>
fun ( char *p)
{
  int i,t; char c[80];
  /**********FOUND**********/
  for (i = 1,t = 0; p[i] ; i++)
    /**********FOUND**********/
    if(!isspace((p+i)))  c[t++]=p[i];
      /**********FOUND**********/
      c[t]="\\0";
  strcpy(p,c);
}
int main( )
{
  char c,s[80];
  int i=0;
  printf("input a string:");
  c=getchar();
  while(c!='#')
  {
    s[i]=c;i++;c=getchar();
  }
  s[i]='\0';
  fun(s);
  puts(s);
  return 0;
}
```

（5）编写一个程序，从键盘输入一个字符串，然后按照字符顺序从小到大进行排序，并删除重复的字符。

```c
#include <stdio.h>
#include <string.h>
int main()
{
  char str[100],*p,*q,*r,c;
  printf("输入字符串:");
  gets(str);
  /**********FOUND**********/
  for(p=str;p;p++)
  {
    for(q=r=p;*q;q++)
      if(*r>*q)
    r=q;
    /**********FOUND**********/
    if(r==p)
    {
      /**********FOUND**********/
      c=r;
      *r=*p;
      *p=c;
    }
  }
  for(p=str;*p;p++)
  {
    for(q=p;*p==*q;q++);
    strcpy(p+1,q);
  }
  printf("结果字符串: %s\n\n",str);
  return 0;
}
```

（6）编写一个函数，求一个字符串的长度，在 main 函数中输入字符串，并输出其长度。

```c
#include <stdio.h>
int length(p)
char *p;
{
  int n;
  n=0;
  /**********FOUND**********/
  while(*p=='\0')
  {
    n++;
    p++;
  }
  return n;
}
int main()
{
```

```
    int len;
    /**********FOUND**********/
    char *str[20];
    printf("please input a string:\n");
    scanf("%s",str);
    /**********FOUND**********/
    len==length(str);
    printf("the string has %d characters.",len);
    return 0;
}
```

3. 设计性实验

（1）实现两个字符串的复制、比较和连接。要求使用指针对字符串操作，不允许调用库函数实现字符串的复制、比较和连接功能。

（2）用动态数组存放 n 个学生某课程的成绩，并计算学生该课程的平均成绩。最后输出所有学生的成绩以及该课程的平均成绩。

（3）编写函数 fun，求一个字符串的长度，在 main 函数中输入字符串，并输出其长度。

```
#include <stdio.h>
int fun(char *p1)
{

    /**********Program**********/




    /**********  End  **********/

}
int main()
{
    char *p,a[20];
    int len;
    p=a;
    printf("please input a string:\n");
    gets(p);
    len=fun(p);
    printf("The string's length is:%d\n",len);
    return 0;
}
```

（4）将主函数中输入的字符串反序存放。例如：输入字符串"abcdefg"，则应输出"gfedcba"。

```
#include <stdio.h>
#include <string.h>
#define N 81
void fun(char *str,int n)
{
```

```
    /**********Program**********/




    /********** End **********/
}
int main()
{
    char s [N];
    int l;
    printf("input a string:");
    gets(s);
    l=strlen(s);
    fun(s,l);
    printf("The new string is :");
    puts(s);
    return 0;
}
```

（5）对长度为 7 个字符的字符串，除首、尾字符外，将其余 5 个字符按降序排列。

例如：原来的字符串为 CEAedca，排序后输出为 CedcEAa。

```
#include<stdio.h>
void fun(char *s,int num)
{
    /**********Program**********/




    /********** End **********/
}
int main()
{
    char s[10];
    printf("输入 7 个字符的字符串:");
    gets(s);
    fun(s,7);
    printf("\n%s",s);
    return 0;
}
```

三、实验注意事项

1. 字符指针变量指向一个字符串常量，只是把该字符串的第 1 个字符的地址赋给了指针变量。
2. 函数名代表该函数的入口地址，指针变量指向函数，表示将函数的入口地址赋给指针变量。

实验 13　结构体

一、实验目的

1. 掌握结构体定义、结构体变量的定义及使用方法。
2. 掌握结构体类型数组的定义、初始化、元素的赋值和使用。
3. 掌握结构体类型变量和共用体类型变量的区别。
4. 了解单链表的建立、访问、插入和删除操作。

二、实验内容

1. 验证性实验

（1）设有 3 名职工的编号、姓名和工资存放在结构体数组中，下面程序输出 3 人中工资居中者的编号、姓名和工资。请填空并调试程序。

```
#include <stdio.h>
struct  worker
{   int num ;
    char name[20] ;
    float salary ;
} workers[ ] ={1, "zhang tian ", 2000, 2, "li ming ", 2400, 3, "wu yu ", 2200} ;
int main()
{
    int  i , j ;
    float max,min;
    max = min = workers[ 0 ] . salary ;
    for ( i = 1; i < 3; i++ )
        if ( workers[ i ]. salary > max )
            _____;
        else if ( workers[ i ].salary < min )
            _____;
    for ( i = 0; i < 3; i++ )
        if ( workers[ i ].salary != max _____ workers[ i ].salary != min)
            printf("%d %c %f\n ",workers[i].num,workers[i].sex, workers[ i ].salary ) ;
    return 0;
}
```

（2）计算某日是当年的第几天。

```
#include <stdio.h>
struct
{
  int year;
  int month;
  int day;
}data;    /* 定义一个结构并声明对象为 data */
int main()
{
    int days;
    printf("请输入日期(年,月,日)：");
    scanf("%d, %d, %d", &data.year, &data.month, &data.day);
```

```
   switch(data.month)
   {
     case 1:days = data.day;
          break;
     /***********SPACE***********/
     case 2:days = data.day+【?】;
          break;
     case 3:days = data.day+59;
          break;
     case 4:days = data.day+90;
          break;
     /***********SPACE***********/
     case 5:days = data.day+【?】;
          break;
     case 6:days = data.day+151;
          break;
     case 7:days = data.day+181;
          break;
     case 8:days = data.day+212;
          break;
     case 9:days = data.day+243;
          break;
     case 10:days = data.day+273;
          break;
     case 11:days = data.day+304;
          break;
     case 12:days = data.day+334;
          break;
   }
   /***********SPACE***********/
   if(data.year%4==0&&data.year%100!=0【?】data.year%400==0)
     if(data.month>=3)
       /***********SPACE***********/
       days = 【?】;
   printf("%d月%d日是%d年的第%d天.\n", data.month, data.day, data.year, days);
   return 0;
 }
```

(3) 设有三人的姓名和年龄保存在结构体数组中，以下程序输出年龄居中者的姓名和年龄。

```
#include <stdio.h>
struct ma
{ char name[20];
  int age;
}person[]={"li", 18, "wang", 19, "zhang", 20};
int main()
{
int i, j, max, min;
 max=min=person[0].age;
  for(i=1; i<3; i++)
/***********SPACE***********/
  if(person[i].age>max)【?】 ;
/***********SPACE***********/
  else if(person[i].age<min)【?】;
 for(i=0; i<3; i++)
/***********SPACE***********/
   if( person[i].age!=max【?】person[i].age!=min)
```

```
           { printf("%s  %d\n", person[i].name, person[i].age);
            break;
            }
    return 0;
}
```

（4）利用指向结构的指针编写求某年、某月、某日是第几天的程序，其中年、月、日和年内天数用结构表示。

```
#include <stdio.h>
#include <stdlib.h>
int main()
{
/***********SPACE***********/
 【?】 date
 {
    int y,m,d,n;
 /***********SPACE***********/
 }【?】;

 int k,f,a[12]={31,28,31,30,31,30,31,31,30,31,30,31};
 printf("date:y,m,d=");
 scanf("%d,%d,%d",&x.y,&x.m,&x.d);
 f=x.y%4==0&&x.y%100!=0||x.y%400==0;
 /***********SPACE***********/
 a[1]+=【?】;
 if(x.m<1||x.m>12||x.d<1||x.d>a[x.m-1]) exit(0);
 for(x.n=x.d,k=0;k<x.m-1;k++)x.n+=a[k];
   /***********SPACE***********/
   printf("n=%d\n",【?】);
   return 0;
}
```

2. 改错性实验

（1）修改编译时发生的语法错误。按照程序中的要求设置断点调试，使得程序的功能是输入 n 个职工的编号、工资，计算其平均工资并输出。

```
#include <stdio.h>
struct worker
{   int num ;
     float salary ;
} workers[10 ];
void input( struct worker * p,int n )
{
    int i ;
    for ( i = 0; i < n; i++ )
    {
        printf("请输入第%d 个职工编号: \n ",i + 1);
        scanf("%d ", ( *p ).num );
        printf("请输入第%d 个职工工资: \n ",i + 1);
        scanf("%f ", p-> num );   //调试时，设置断点
    }
}
float average ( struct worker * p,int n )
{
        int i ;
```

```
        double sum;
        for ( i = 0; i < n; i++ )
            sum + = p->salary ;
        return sum,n ;   //调试时,设置断点
}
int main()
{
        struct worker * p = workers ;
        int  n ;
        printf("请输入职工人数,不超过10人: ") ;
        scanf("%d " , &n );   //调试时,设置断点
        input(* p, n );
        printf("%d 个职工的平均工资为: %10.2f ",n, average(p, n) ) ;   //设置断点
        return 0;
}
```

（2）下题是一段有关结构体变量传递的程序。

```
#include <stdio.h>
struct student
{
  int x;
  char c;
} a;
f(struct student b)
{
  b.x=20;
  /**********FOUND**********/
  b.c=y;
  printf("%d,%c",b.x,b.c);
}
int main()
{
  a.x=3;
  /**********FOUND**********/
  a.c='a'
  f(a);
  /**********FOUND**********/
  printf("%d,%c",a.x,b.c);
  return 0;
}
```

3. 设计性实验

（1）编写一个 C 程序，定义一个日期结构变量 date（由年 year、月 month、日 day 3 个整型数据组成），从键盘为该变量中的各成员输入数据，然后再将输入的日期显示出来。

```
#include <stdio.h>
int main()
{

/**********Program**********/


/********** End **********/
return 0;
}
```

（2）书店新进了5门教材，输入其书名、册数、单价，按它们的册数降序排序，并求各门教材的金额。

（3）学生记录由学号、姓名、总分构成，通过键盘为 N 名学生输入数据，存于结构体数组中。编写函数 fun，找出成绩最低的学生记录，通过形参返回（规定只有一个最低分）。

（4）学生记录由学号、姓名、总分构成，通过键盘为 N 名学生输入数据，存于结构体数组中。编写函数 fun，把高于平均分的学生数据放在 b 所指的数组中，高于平均分的学生人数通过形参 n 返回，平均分通过函数值返回。

三、实验注意事项

1. 注意声明结构体类型的形式，结尾处的分号不能省略。
2. 注意区别结构体类型的声明和结构体变量的定义。

实验 14　文件

一、实验目的

1. 掌握文件的概念，结构体类型 FILE。
2. 了解文件操作的基本步骤以及错误处理。
3. 掌握文件操作相关函数，如文件的打开和关闭、文件读写函数等。

二、实验内容

1. 验证性实验

（1）从键盘输入若干行（每行不超过 80 个字符），写到文件 myfile1.txt 中，用 -1 作为字符串输入结束的标志。然后将文件内容显示在屏幕上。文件的读写分别由自定义函数 readText 和 writeText 实现。程序填空。

```
#include <stdio.h>
#include <string.h>
#include <stdliB.h>
void writeText(FILE *);
void readText(FILE *);
int main()
{
    FILE * fp;
    if( (fp = fopen("myfile1.txt ", "w "))== NULL )
        {   printf("文件打开失败!\n "); exit(0); }
    writeText(fp);
    fclose(fp);
    if( (fp = fopen("myfile1.txt ", "r "))== NULL )
        {   printf("文件打开失败!\n "); exit(0); }
    readText(fp);
    fclose(fp);
    return 0;
}
void writeText(FILE _____ )
```

```
    {
        char str[81];
        printf("\n输入一个字符串，以-1结尾：\n ");
        gets(str);
        while( strcmp(str,"-1 ")!=0 )
        {
            fputs(_____,fw);
            fputs("\n ",fw );
            gets( str );
        }
    }
    void readText(FILE *fr )
    {
            char str[81];
            printf("\n读文件并输出到屏幕\n ");
            fgets(str, 81,fr );
            while( !feof( fr ))
            {
                printf("%s ", _____);
                fgets(str, 81,fr );
            }
    }
```

（2）以下程序中，函数 fun 的功能是：将自然数 1～10 以及它们的平方根写到文件 myfile2.txt 中，然后再按顺序读出显示到屏幕上。

```
#include <stdio.h>
#include <math.h>
int fun(char *fname )
{
    FILE * fp; int i,n; float x;
    if( (fp = fopen("fname ", "w "))== NULL )  return 0;
    for( i=1; i<= 10; i++)
        fprintf(_____, "%d %f\n ",I,sqrt((double)i ));
    printf("\n写入成功!\n ");
    _____;
    printf("\n文件中的数据：\n ");
    if( (fp = fopen(_____, "r "))== NULL )  return 0;
    fscanf( fp,"%d%f\n ",&n,&x );
    while( ! feof(fp))
    {
        printf("%d %f\n ", n, x);
        fscanf( fp,"%d%f\n ",&n,&x );
    }
    fclose(fp); return 1;
}
int main( )
{
    char fname[]="myfile3.txt ";
    fun( fname );
    return 0;
}
```

（3）题目：以下程序打开文件后，先利用 fseek 函数将文件位置指针定位在文件末尾，然后调用 ftell 函数返回当前文件位置指针的具体位置，从而确定文件长度。

```
#include <stdio.h>
int main()
{
      FILE *myf;
      long f1;
/***********SPACE***********/
      myf= 【?】("test.t","rb");
/***********SPACE***********/
      fseek (myf, 0, 【?】 );
      f1=ftell (myf);
      fclose (myf);
      printf ("%d\n",f1);
      return 0;
}
```

（4）说明：下面程序的功能是统计文件中字符个数，请填写程序所缺内容。

```
#include "stdio.h"
int main()
{
  /***********SPACE***********/
  【?】*fp;
  long num=0L;
     if((fp=fopen("fname.dat","r"))==NULL)
      {
         printf("Open error\n");
      }
  /***********SPACE***********/
     while(【?】)
       {
         fgetc(fp);
         num++;
       }
  /***********SPACE***********/
     printf("num=%1d\n",【?】);
     fclose(fp);
     return 0;
}
```

（5）功能：从键盘输入一个字符串，将小写字母全部转换成大写字母，然后输出到一个磁盘文件"test"中保存。输入的字符串以！结束。

```
#include <stdio.h>
#include <string.h>
#include <stdlib.h>

int main()
{
  FILE *fp;
  char str[100];
  int i=0;
  /***********SPACE***********/
  if((fp=fopen("test",【?】))==NULL)
  {
    printf("cannot open the file\n");
    exit(0);
```

```
    }
    printf("please input a string:\n");
    /**********SPACE**********/
    gets(【?】);
    while(str[i]!='!')
    {
    /**********SPACE**********/
      if(str[i]>='a'&&【?】)
        str[i]=str[i]-32;
      fputc(str[i],fp);
      i++;
    }
    /**********SPACE**********/
    fclose(【?】);
    fp=fopen("test","r");
    fgets(str,strlen(str)+1,fp);
    printf("%s\n",str);
    fclose(fp);
    return 0;
}
```

2. 改错性实验

（1）功能：将若干学生的档案存放在一个文件中，并显示其内容。

```
#include <stdio.h>
struct student
{
  int num;
  char name[10];
  int age;
};
struct student stu[3]={{001,"Li Mei",18},
                       {002,"Ji Hua",19},
                       {003,"Sun Hao",18}};
int main()
{
  /**********FOUND**********/
  struct student p;
  /**********FOUND**********/
  cfile fp;
  int i;
  if((fp=fopen("stu_list","wb"))==NULL)
  {
    printf("cannot open file\n");
    return;
  }
  /**********FOUND**********/
  for(*p=stu;p<stu+3;p++)
    fwrite(p,sizeof(struct student),1,fp);
  fclose(fp);
  fp=fopen("stu_list","rb");
  printf(" No.   Name      age\n");
  for(i=1;i<=3;i++)
  {
    fread(p,sizeof(struct student),1,fp);
    /**********FOUND**********/
    scanf("%4d %-10s %4d\n",*p.num,p->name,(*p).age);
  }
```

```
        fclose(fp);
        return 0;
}
```

（2）功能：将一组职工信息写到文件 data.txt 中。

```
#include <stdio.h>
#define N 3
struct  datatype{
        long int id;
        char name[20];
        struct {
                int y,m,d;
        }workdate;
};
int main(){
        struct  datatype d[N]={{10001,"李四",{2000,6,7}}
            ,{10003,"赵六",{2002,7,7}},{10008,"王五",{2002,7,7}}};
        int i;
/**********FOUND**********/
  file *f
  f=fopen("data.txt","w");
  for(i=0;i<N;i++)
/**********FOUND**********/
        fprintf("%ld %s %d %d %d \n",d[i].id,d[i].name,
/**********FOUND**********/
            d[i].y,d[i].m,d[i].d);
/**********FOUND**********/
  fclose(data);
  return 0;
}
```

3. 设计性实验

（1）编写程序，删除一个 C 语言源程序中的所有注释信息，并将删除的注释内容写入一个文件中。

（2）从键盘输入若干行字符，输入后把它们存储到一磁盘文件中。在从该文件中读出这些数据，将其中的小写字母转换成大写字母后在屏幕上输出。

（3）编写程序，建立并显示同学录。要求将 5 名同学的信息（姓名、性别、出生日期、家庭住址、电话、QQ 号）放在一个结构体数组中，这些信息由键盘输入并存储到磁盘文件 myfile3.txt 中。再从该文件中读出同学通讯录信息，按以下格式输出到屏幕上。

姓　名	性别	出生日期	家庭住址	电话	QQ
李一飞	男	1990-10-10	山东泰安	12345678900	123456789
……					

三、实验注意事项

1. C 语言文件的使用步骤：程序中首先包含头文件#include<stdio.h>；定义文件类型指针（FILE *fp;）；打开文件；读写文件；关闭文件。

2. 打开文件 fopen 函数中，注意文件的使用方式。

3. 注意 fscanf 函数和 scanf 函数的区别，以及 fprintf 函数和 printf 函数的区别。

实验 15　链表

一、实验目的

1. 理解单链表的基本概念。
2. 掌握单链表的建立、访问、插入和删除等操作。

二、实验内容

设计性实验

（1）采用表尾插入的方法建立一个包含 10 个学生的链表。学生信息包括：学号、姓名、数学成绩、英语成绩和总分。

（2）编写 5 个函数，它们的功能分别为：建立含 n 个整数结点的单链表 list、统计 list 表中的结点数、输出 list 表中的数据、在表中指定位置之后插入一个整数结点、在表中指定位置之前删除一个结点。编写 main 函数时调用上述函数。

三、实验注意事项

链表中每个节点都应包含两个部分：实际数据和下一个节点的地址。通常，链表的"头指针"变量以 head 表示，存放于一个地址，该地址指向一个节点。链表的"表尾"不再指向其他节点，其地址部分放一个"NULL"，链表至此结束。

第3部分 综合性实验

本部分为读者设置了3个综合性实验,读者可根据需要选择。这3个实验项目都给出了参考程序,希望读者首先独立思考、独立编程,然后再和书中给出的答案进行比较,考察两者的优劣。通过综合性实验,让学生熟练使用结构化程序的设计思想和方法,培养学生综合应用知识的能力。

实验1 学生成绩管理

一、实验要求

编辑一个可执行的学生成绩管理系统,图3-1所示为该系统的功能模块结构图。

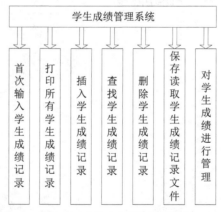

图3-1 系统功能模块结构图

二、实验提示

用结构体定义学生的成绩记录单,包括:学生的姓名、学号、性别、年龄、四科成绩(大学英语、高等数学、C语言、马克思主义哲学原理)、四科成绩的总分和平均分。

对成绩单的存储采用单链表进行,每个结点存储一个学生记录。对成绩记录的各种操作都基于建立的链表进行:链表的建立对应成绩记录的首次录入,链表的删除对应成绩记录的删除,链表的插入对应于成绩记录的插入,链表的查询对应成绩记录的查询,链表的输出对应成绩记录的打印。

使用链表作为存储结构,既避免了像数组那样需要开辟很大的连续的存储单元,又加快了成绩记录在程序中的计算速度。链表的各种操作并不对其他结点造成影响。

结构体的变量类型：姓名采用字符串（如罗欢）、性别为单字符（m/f）、学号为字符串（如52092202）、年龄和成绩采用整形变量、平均分用浮点型。

```
typedef struct student
{
    char num[12];       //学号
    char name[15];      //姓名
    int age;            //年龄
    char sex;           //性别
    int score[4];       //成绩
    float ave;          //平均成绩
    float sum;          //总成绩
    int order;          //排名
    struct student *next;//指向下一结点的指针
}stu;
```

下面给出其中几个函数块的N-S图。

（1）创建函数，如图3-2所示。

（2）查找函数，如图3-3所示。

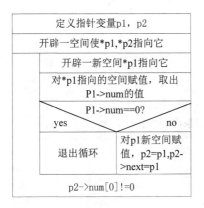

图3-2 创建函数

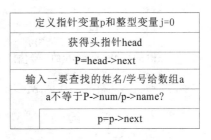

图3-3 查找函数

（3）插入函数，如图3-4所示。

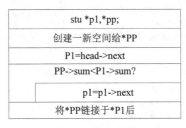

图3-4 插入函数

三、参考程序

```c
#include<stdio.h>
#include<stdlib.h>
```

```c
#include<time.h>
#include<string.h>
typedef struct student   //定义学生变量的结构体类型
{
    char num[12];
    char name[15];
    int age;
    char sex;
    int score[4];
    float ave;
    float sum;
    int order;
    struct student *next;
}stu;
int n=0;
int tui=1;
float g=0;
#define LONG sizeof(stu)
stu *creat()       //创建学生信息管理系统的初始记录
{
    stu *head,*p1,*p2;
    int i;
    float s=0;
    p1=p2=(stu*)malloc(LONG);
    head=NULL;
    do
    {
        n++;
        if(n==1)head=p1;
        p1=(stu*)malloc(LONG);
        printf("输入该学生的学号[输入x退出成绩登录]:\n");
        scanf("%s",p1->num);
        if(p1->num[0]=='x'||p1->num[0]=='X')
        {
            p2->next=NULL;
            return(head);
        }
        printf("输入该学生姓名:\n");
        scanf("%s",p1->name);
        printf("输入该学生年龄:\n");
        scanf("%d",&p1->age);
        printf("输入该学生性别[m for man,f for women]:\n");
        getchar();
        p1->sex=getchar();
        for(i=1;i<=4;i++)
        {
            printf("输入第 %d 科 成绩:\n",i);
            scanf("%d",&p1->score[i-1]);
            while((p1->score[i-1]>100)||(p1->score[i-1]<0))
            {
                printf("输入错误请再次输入:\n");
                printf("输入第 %d 科 成绩:\n",i);
                scanf("%d",&p1->score[i-1]);
            }
```

```c
                s=s+p1->score[i-1];
            }
            p1->ave=s/4;
            p1->sum=s;
            p2->next=p1;
            p2=p1;
        }while(p2->num[0]!=0);
        return(head);
    }
    void print(stu *head)      //输出学生管理系统中的记录
    {
        system("cls");
        stu *p;
        int i=0,j=0;
        float s=0;
        char k;
        p=head->next;
        while(p!=NULL)
        {
            s=0;
            j++;
            k=p->sex;
            printf("第 %2d 个学生记录:\n",j);
            printf("姓名:%12s\t学号:%10s\t",p->name,p->num);
            if(k=='m')
            {
                printf("性别:男\t");
            }
            else
            {
                printf("性别:女\t");
            }
            printf(" 排 名 :%4d\n 大学 英语 :%3d\t 高 数 :%3d\tC 语 言 :%3d\t 马
哲:%3d\n",p->order,p->score[0],p->score[1],p->score[2],p->score[3]);
            for(i=0;i<4;i++)
            {
                s=s+p->score[i];
            }
            p->sum=s;
            p->ave=s/4;
            printf("总分: %5.2f\t 平均分:%5.2f\n",p->sum,p->ave);
            p=p->next;
        }
        system("pause");
    }
    stu *insert(stu *head)      //插入记录
    {
        system("cls");
        void sort(stu *head);
        stu *p1,*pp;
        int i=0;
        float s=0;
        pp=(stu *)malloc(LONG);
        printf("输入该学生的学号:\n");
```

```c
        scanf("%s",pp->num);
        printf("输入该学生姓名:\n");
        scanf("%s",pp->name);
        printf("输入该学生年龄: \n");
        scanf("%d",&pp->age);
        printf("输入该学生性别[m for man,f for women]:\n");
        getchar();
        pp->sex=getchar();
        printf("1.大学语\t2.高数\t3.C语言\t4.马哲\n");
        for(i=1;i<=4;i++)
        {
            printf("输入第 %d 科 成绩:\n",i);
            scanf("%d",&pp->score[i-1]);
            while((pp->score[i-1]>100)||(pp->score[i-1]<0))
            {
                printf("输入错误请再次输入:\n");
                printf("输入第 %d 科 成绩:\n",i);
              scanf("%d",&pp->score[i-1]);
            }
            s=s+pp->score[i-1];
        }
        pp->ave=s/4;
        pp->sum=s;
        p1=head->next;
        while(p1->next!=NULL)
        {
            p1=p1->next;
        }
        if(p1->next==NULL)
        {
            p1->next=pp;
            pp->next=NULL;
        }
        sort(head);
        n=n+1;
        return(head);
}
void search(stu *head)      //查找记录
{
    stu *p;
    int i=1,j=0;
    char a[15],k;
    p=head->next;
    printf("本次查找[1.基于学生姓名 2.基于学生学号]:\n");
    scanf("%d",&i);
    if(i==1)
    {
        printf("输入要查询的学生姓名:\n");
        scanf("%s",a);
        while((p!=NULL)&&(j==0))
        {
            k=p->sex;
            if(strcmp(a,p->name)==0)
            {
```

```c
                printf("查找成功！\n查找结果:\n");
                printf("姓名:%12s\t学号:%10s\t",p->name,p->num);
                if(k=='m')    printf("性别:男\t");
                else          printf("性别:女\t");
                printf("排 名:%4d\n 大 学 英 语:%3d\t 高 数:%3d\tC 语 言:%3d\t 马 哲:%3d\n",p->order,p->score[0],p->score[1],p->score[2],p->score[3]);
                j=1;
            }
            else    p=p->next;
        }
        if(p==NULL)printf("抱歉！该系统内无该同学成绩记录。");
        system("pause");
    }
    else
    {
        printf("输入要查询的学生学号:\n");
        scanf("%s",a);
        while((p!=NULL)&&(j==0))
        {
            k=p->sex;
            if(strcmp(a,p->num)==0)
            {
                printf("查找成功！\n查找结果:\n");
                printf("姓名:%12s\t学号:%10s\t",p->name,p->num);
                if(k=='m')    printf("性别:男\t");
                else          printf("性别:女\t");
                printf("排 名:%4d\n 大 学 英 语:%3d\t 高 数:%3d\tC 语 言:%3d\t 马 哲:%3d\n",p->order,p->score[0],p->score[1],p->score[2],p->score[3]);
                j=1;
            }
            else    p=p->next;
        }
        if(p==NULL)printf("抱歉！该系统内无该同学成绩记录。");
        system("pause");
    }
}
void sort(stu *head)    //记录排序
{
    stu *p,*q,*r;
    int i=0;
    p=head->next;
    if(p!=NULL)
    {
        r=p->next;
        p->next=NULL;
        p=r;
        while(p!=NULL)
        {
            r=p->next;
            q=head;
            while((q->next!=NULL)&&(q->next->sum>p->sum))q=q->next;
            p->next=q->next;
            q->next=p;
```

```c
                p=r;
            }
        }
        p=head->next;
        i=1;
        while(p!=NULL)
        {
            p->order=i;
            p=p->next;
            i++;
        }
        printf("排序成功! 选择打印查看\n");
        system("pause");
}
void savefile( stu *head)         //保存文件
{
        FILE *fp;
        stu *p0;
        char filename[12];
        p0=head->next;
        system("cls");
        printf("请输入保存文件名:");
        scanf("%s",filename);
        if((fp=fopen(filename,"wb"))==NULL)
          {
            printf("无法打开\n");
            return;
          }
        while(p0!=NULL)
        {
          if(fwrite(p0,LONG,1,fp)!=1)
          printf("文件写入错误\n");
          p0=p0->next;
        }
        fclose(fp);
}
stu *loadfile()         //读文件
{
        FILE *fp;
        stu *p0,*p1,*head;
        char filename[12];
        n=0;
        system("cls");
        printf("请输入数据源文件名:");
        scanf("%s",filename);
        head=(stu*)malloc(LONG);
        if((fp=fopen(filename,"rb"))==NULL)
        {
                printf("文件无法打开\n");
                return head;
        }
        head->next=NULL;
        p0=(stu*)malloc(LONG);
```

```c
            if(fread(p0,LONG,1,fp)!=1)
              {
                    printf("读入错误\n");
                    return head;
              }
              else
              {
                    head->next=p1=p0;
                    n=1;
              }
            while(!feof(fp))
            {
                p0=(student*)malloc(LONG);
                if(fread(p0,LONG,1,fp)!=1)
                  {
                        printf("file write error\n");
                        return head;
                  }
                p1->next=p0;
                 p1=p0;
                n++;
            }
            fclose(fp);
            return(head);
}
void delet(stu *head)       //删除记录
{
      int a=1,c=0;
      char b[15],k;
      stu *h,*p;
      h=head->next;
      p=head;
      printf("选择 1.基于姓名删除 2.基于学号删除:\n");
      scanf("%d",&a);
      if(a==1)
      {
            printf("输入您要删除记录姓名:\n");
            scanf("%s",b);
            while(h!=NULL)
            {
                  if(strcmp(b,h->name)==0)
                  {
                        k=h->sex;
                        printf("姓名:%12s\t学号:%10s\t",h->name,h->num);
                        if(k=='m')    printf("性别:男\t");
                        else        printf("性别:女\t");
                        printf(" 排 名 :%4d\n 大学英语 :%3d\t 高数 :%3d\tC 语言 :%3d\t 马
哲:%3d\n",h->order,h->score[0],h->score[1],h->score[2],h->score[3]);
                        printf("是否删除? [1.是 2.否]\n");
                        scanf("%d",&c);
                        if(c==1)
                        {
                              if(h->next==NULL)
                              {
```

```c
                    p->next=NULL;free(h);
                    h=NULL;
                }
                else
                {
                    p->next=h->next;free(h);
                    h=NULL;
                }
                n--;
            }
            else    return;
        }
        else
        {
            p=h;
            h=h->next;
        }
    }
}
else
{
    printf("输入您要删除的记录学号:\n");
    scanf("%s",b);
    while(h!=NULL)
    {
        if(strcmp(b,h->num)==0)
        {
            k=h->sex;
            printf("姓名:%12s\t学号:%10s\t",h->name,h->num);
            if(k=='m')    printf("性别:男\t");
            else          printf("性别:女\t");
            printf(" 排 名 :%4d\n 大 学 英 语 :%3d\t 高 数 :%3d\tC 语 言 :%3d\t 马 哲:%3d\n",h->order,h->score[0],h->score[1],h->score[2],h->score[3]);
            printf("有该学生记录是否删除? [1.是 2.否]");
            scanf("%d",&c);
            if(c==1)
            {
                if(h->next==NULL)
                {
                    p->next=NULL;free(h);
                    h=NULL;
                }
                else
                {
                    p->next=h->next;free(h);
                    h=NULL;
                }
                n--;
            }
            else    return;
        }
        else
        {
            p=h;
```

```c
                    h=h->next;
                }
            }
        }
    }
    void quit()        //退出系统
    {
        int a;
        printf("是否退出系统:[1.是 2.否]\n");
        scanf("%d",&a);
        if(a==1)
        {
                printf("谢谢您的使用!!! 系统立即为你退出...");
                exit(0);
        }
        else
        {
            printf("返回界面...");system("pause");
        }
    }
    int jiemian()       //主界面
    {
        struct tm *pt;
        int a;
        time_t t;
        t=time(NULL);
        pt=localtime(&t);
        system("cls");
        printf("~~~~~~~~~~~~~~~~~~学生成绩登录系统 欢迎您!~~~~~~~~~~~~~~~~~~\n");
        printf("选择您要执行的操作: ");
printf("\n1.首次输入学生成绩记录\n2.对首次输入的记录按成绩排序：\n3.打印所有学生记录\n4.按顺序插入学生记录\n5.查找学生成绩记录\n6.删除学生成绩记录\n7.保存记录\n8.载入已保存的记录\n9.成绩整体统计\n10.退出系统\n");
        scanf("%d",&a);
        return(a);
    }
    void zhengli(stu *head)      //记录整理
    {
        int j=2,i,k=0;
        int a=0,b=0,c=0,d=0,e=0;
        float s=0;
        stu *p;
        p=head->next;
        printf("学生记录统计: \n");
        printf("是否打印不及格学生记录[1.是2.否]\n");
        scanf("%d",&j);
        for(i=1;i<=4;i++)
        {
            p=head->next;
            a=0;b=0;c=0;d=0;e=0;s=0,k=0;
            while(p!=NULL)
            {
                k++;
```

```
                if(i==1)g=g+p->sum;
                s=s+(float)p->score[i-1];
                if(p->score[i-1]>=90)a++;
                else if(p->score[i-1]>=80)b++;
                else if(p->score[i-1]>=70)c++;
                else if(p->score[i-1]>=60)d++;
                else
                {
                    e++;
                    if(j==1)
                    {
                    printf("排名第 %3d 学生 第 %3d 科有不及格记录：\n 分数为：
                    %2d\n",p->order,i,p->score[i-1]);
                    }
                }
                p=p->next;
            }
            if(i==1)g=g/(float)k;
            s=s/(float)k;
            printf("第%2d 科成绩\t 平均分:%5.2f\t 90~100 分%2d 人 80~90 分%2d 人   70~80 分%2d 人
60~70 分%2d 人 不及格%2d 人\n",i,s,a,b,c,d,e);
        }
        printf("总成绩平均分：\t%5.2f\t 总人数:%4d\n",g,n);
}

void main()
{
    stu *head;
    while(tui==1)
    {
        switch(jiemian())
        {
        case 1:head=creat();break;
        case 2:sort(head);break;
        case 3:print(head);break;
        case 4:head=insert(head);break;
        case 5:search(head);break;
        case 6:delet(head);break;
        case 7:savefile(head);break;
        case 8:head=loadfile();break;
        case 9:zhengli(head);
        case 10:quit();
        }
    }
}
```

实验2 约瑟夫环问题

一、实验要求

约瑟夫（Josephus）环问题的一种描述是：编号为 1，2，…，n 点的 n 个人按顺时针方向围坐一个圈,每人持有一个密码。一开始选一个正整数作为报数上限值 m,从第一个人开始按顺时针方向自 1 开始报数,报到 m 时停止。报到 m 的人出列,将他的密码作为新的 m 值,从他在顺时针方向上的下一个人开始重新报数,如此下去,直到所有人出列。

基本要求:利用单向循环链表存储结构模拟此过程,按照出列的顺序输出每个人的编号。

测试数据:m 的初始值为 20;n=7,7 个人的密码依次是 3,1,7,2,4,8,4; m 的初始值为 6(正确的出列顺序为 6,1,4,7,2,3,5)。

二、实验提示

采用结构体定义单链表,格式为:

```
struct Lnode{
    int number;           //序号
    int password;         //密码
    struct Lnode *next;   //指向下一结点的指针
}Lnode,*p,*q,*head;
```

注:Lnode 是结点变量,p、q 是结点,head 是头指针。

三、参考程序

```
#include<stdio.h>
#include<stdliB.h>
struct Lnode                    //定义链表
{int number;
 int password;
 struct Lnode *next;
}Lnode,*p,*q,*head;

int main()
{int n;                //n 个人
 int i;
 int m;                //初始报数上限值
 int j;
 printf("please enter the number of people n:");        //输入测试人的数量
 scanf("%d",&n);
 loop:if(n<=0||n>30)            //检验 n 是否满足要求,如不满足重新输入 n 值
  {printf("\n  n is erorr!\n\n");
   printf("please enter the number of people again n:");
   scanf("%d",&n);
   goto loop;
   }
 for(i=1;i<=n;i++)              //建立单链表
  {if(i==1)
    {head=p=(struct Lnode*)malloc(sizeof(struct Lnode)); }
   else
    {q=(struct Lnode*)malloc(sizeof(struct Lnode));
     p->next=q;
     p=q;
     }
    printf("please enter the %d people's password:",i);    //输入每个人所持有的密码值
```

```
        scanf("%d",&(p->password));
        p->number=i;
      }
    p->next=head;                              //形成循环链表
    p=head;
    printf("please enter the number m:");
    scanf("%d",&m);
    printf("The password is:\n");
    for (j=1;j<=n;j++)                         //输出每个人的编号
    {for(i=1;i<m;i++,p=p->next);
      m=p->password;
      printf("%d  ",p->number);
      p->number=p->next->number;               //删除报m的结点
      p->password=p->next->password;
      q=p->next;
      p->next=p->next->next;
      free(q);
      }
    printf("\n\n");
}
```

实验3　双向链表的综合运算

一、实验要求

构建一个双向链表，实现插入、查找和删除操作。

二、实验提示

在双向链表中，结点除含有数据域外，还有两个链域。一个存储直接后继结点地址，一般称为右链域；一个存储直接前驱结点地址，一般称为左链域。双向链表结点类型可以定义为：

```
typedef struct node
{
int data;   //数据域
struct node *llink,*rlink;  //链域，*llink是左链域指针，*rlink是右链域指针
}JD;
```

三、参考程序

```
#include <stdio.h>
#include <malloc.h>
#include <string.h>
#include <stdlib.h>
#define N 10

typedef struct node
{
    char name[20];
```

```c
    struct node *llink,*rlink;
}stud;

stud * creat(int n)      //生成双向循环链表
{
    stud *p,*h,*s;
    int i;
    if((h=(stud *)malloc(sizeof(stud)))==NULL)
    {
        printf("不能分配内存空间!");
        exit(0);
    }
    h->name[0]='\0';
    h->llink=NULL;
    h->rlink=NULL;
    p=h;
    for(i=0;i<n;i++)
    {
        if((s= (stud *) malloc(sizeof(stud)))==NULL)
        {
            printf("不能分配内存空间!");
            exit(0);
        }
        p->rlink=s;
        printf("请输入第%d 个人的姓名",i+1);
        scanf("%s",s->name);
        s->llink=p;
        s->rlink=NULL;
        p=s;
    }
    h->llink=s;
    p->rlink=h;
    return(h);
}

void search(stud *h)     //查找链表中某个结点
{
    int k,i=1;
    stud *p;
    p=h->rlink;
    printf("输入要查找结点的序号：\n");
    scanf("%d",&k);
    while((p!=h)&&(i<k))
    {
        p=p->rlink;
        i++;
    }
    if(i>=k) printf("%s",p->name);
    else    printf("没找到该结点!\n");
}

void print(stud *h)      //输出链表中所有结点的信息
{
    stud *p;
```

```c
        p=h->rlink;
        printf("数据信息为：\n");
        while(p!=h)
        {
            printf("%s ",&*(p->name));
            p=p->rlink;
        }
        printf("\n");
}
stud *insert(stud *head,int i)      //在链表中第i个结点之间插入一个新结点
{
        char stuname[20];
        stud *s,*p;
        int j;
        p=head ;j=0;
        while((p->rlink!=head)&&(j<i-1))
        {
            p=p->rlink;
            j++;
        }
        if(j==i-1){
            if((s= (stud *) malloc(sizeof(stud)))==NULL)
            {
                printf("不能分配内存空间!");
                exit(0);
            }
            printf("请输入你要插入的人的姓名:");
            scanf("%s",stuname);
            strcpy(s->name,stuname);
            s->rlink=p->rlink;
            p->rlink=s;
            s->llink=p;
            (s->rlink)->llink=s;
        }
        else    printf("error\n");
        return head;
}
stud *del(stud *head,int i)      //删除链表中第i个结点
{
        stud *p;
        int j;
        p=head;
        j=0;
        while((p->rlink!=head)&&(j<i))
        {
            p=p->rlink;
            j++;
        }
        if(j==i){
            (p->rlink)->llink=p->llink;
            (p->llink)->rlink=p->rlink;
            free(p);
        }
        else    printf("error\n");
        return head;
```

```c
}
void main()
{
    int number,flag=1,i,address;
    stud *head;
    while(flag){
        printf("\n1.建立双向循环链表\n");
        printf("2.查找某个结点\n");
        printf("3.插入一个结点\n");
        printf("4.删除一个结点\n");
        printf("5.退出\n");
        printf("请输入选择：\n");
        scanf("%d",&i);
        switch(i){
        case 1:number=N;
            head=creat(number);
            print(head);
            break;
        case 2:   search(head);break;
        case 3:
            printf("输入待插入结点的序号：\n");
            scanf("%d",&address);
            head=insert(head,address);
            print(head);
            break;
        case 4: printf("输入待删除结点的序号：\n");
            scanf("%d",&address);
            head=del(head,address);
            print(head);
            break;
        case 5: flag=0;
        }// end switch
    }//end while
}//end main
```

第4部分 提高性实验

提高性实验 1

一、填空型实验

（1）功能：从键盘上输入一个字符串，将该字符串升序排列后输出到文件 test.txt 中，然后从该文件读出字符串并显示出来。

```
#include<stdio.h>
#include<string.h>
#include<stdlib.h>
int main()
{
   FILE  *fp;
   char t,str[100],str1[100];    int n,i,j;
   if((fp=fopen("test.txt","w"))==NULL)
   {
      printf("can't open this file.\n");
      exit(0);
   }
   printf("input a string:\n"); gets(str);
   /***********SPACE***********/
   【?】;
   /***********SPACE***********/
   for(i=0; 【?】 ;i++)
     for(j=0;j<n-i-1;j++)
     /***********SPACE***********/
      if(【?】)
      {
         t=str[j];
         str[j]=str[j+1];
         str[j+1]=t;
      }
   /***********SPACE***********/
   【?】;
   fclose(fp);
   fp=fopen("test.txt","r");
```

```
    fgets(str1,100,fp);
    printf("%s\n",str1);
    fclose(fp);
    return 0;
}
```

（2）功能：调用函数 fun 计算 m=1-2+3-4+…+9-10，并输出结果。

```
#include <stdio.h>
int fun( int n)
{
  int m=0,f=1,i;
  /**********SPACE**********/
  for(i=1;【?】;i++)
  {
    m+=i*f;
    /**********SPACE**********/
    【?】;
  }
  /**********SPACE**********/
  return 【?】;
}
int main()
{
  printf("m=%d\n", fun(10));
  return 0;
}
```

二、改错型实验

（1）功能：根据整型形参 m 的值，计算如下公式的值。

$$t = 1 - \frac{1}{2\times 2} - \frac{1}{3\times 3} - \cdots - \frac{1}{m\times m}$$

例如：若 m=5，则应输出 0.536389。

```
#include <stdio.h>
double fun(int m)
{
  double y=1.0;
  int i;
  /**********FOUND**********/
  for(i=2;i<m;i--)
    /**********FOUND**********/
    y-=1/(i*i);
  /**********FOUND**********/
  return m;
}
int main()
{
  int n=5;
  printf("\nthe result is %lf\n",fun(n));
  return 0;
}
```

（2）功能：一个 5 位数，判断它是不是回文数。例 12321 是回文数，个位（最低位）与万位（最高位）相同，十位与千位相同。

```
#include<stdio.h>
int main( )
{
  /**********FOUND**********/
  long ge,shi,qian;wan,x;
  scanf("%ld",&x);
  /**********FOUND**********/
  wan=x%10000;
  qian=x%10000/1000;
  shi=x%100/10;
  ge=x%10;
  /**********FOUND**********/
  if (ge==wan||shi==qian)
    printf("this number is a huiwen\n");
  else
    printf("this number is not a huiwen\n");
  return 0;
}
```

（3）功能：编写函数 fun，求 20 以内所有 5 的倍数之积。

```
#define N 20
#include <stdio.h>
int fun(int m)
{
  /**********FOUND**********/
  int s=0,i;
  for(i=1;i<N;i++)
    /**********FOUND**********/
    if(i%m=0)
      /**********FOUND**********/
      s=*i;
  return s;
}
int main()
{
  int sum;
  sum=fun(5);
  printf("%d 以内所有%d 的倍数之积为： %d\n",N,5,sum);
  return 0;
}
```

三、设计型实验

（1）功能：编写函数求 3!+6!+9!+12!+15!+18!+21!的值。

```
#include <stdio.h>
float sum(int n)
{
  /**********Program**********/

  /********** End **********/
}
```

```
int main()
{
  printf("this sum=%e\n",sum(21));
  return 0;
}
```

（2）功能：删除所有值为 y 的元素。数组元素中的值和 y 的值由主函数通过键盘输入。

```
#include <stdio.h>
#include<conio.h>
#include<stdio.h>
#define M 20
void fun(int bb[],int *n,int y)
{
  /**********Program**********/

  /********** End **********/
}
int main()
{
  int aa[M],n,y,k;
  printf("\nPlease enter n:");scanf("%d",&n);
  printf("\nEnter %d positive number:\n",n);
  for(k=0;k<n;k++) scanf("%d",&aa[k]);
  printf("The original data is:\n");
  for(k=0;k<n;k++) printf("%5d",aa[k]);
  printf("\nEnter a number to deletede:");scanf("%d",&y);
  fun(aa,&n,y);
  printf("The data after deleted %d:\n",y);
  for(k=0;k<n;k++) printf("%4d",aa[k]);
  printf("\n");
  return 0;
}
```

提高性实验 2

一、填空型实验

（1）功能：删除字符串中的指定字符，字符串和要删除的字符均由键盘输入。

```
#include <stdio.h>
int main()
{
  char str[80],ch;
  int i,k=0;
  /***********SPACE***********/
  gets(【?】);
  ch=getchar();
  /***********SPACE***********/
  for(i=0;【?】;i++)
    if(str[i]!=ch)
    {
```

```
    /**********SPACE**********/
      【?】;
      k++;
    }
  /**********SPACE**********/
  【?】;
   puts(str);
  return 0;
}
```

(2)功能：以下程序的功能如下。

```
pi=1-1/3+1/5-1/7+1/9-1/11+……
#include <stdio.h>
#include <math.h>
int main()
{
  int f;
  /**********SPACE**********/
  【?】;
  double t,pi;
  t=1;pi=t;f=1;n=1.0;
  /**********SPACE**********/
  while(【?】 )
  {
    n=n+2;
    /**********SPACE**********/
    【?】;
    t=f/n;
    pi=pi+t;
  }
  /**********SPACE**********/
  【?】;
   printf("pi=%10.6f\n",pi);
  return 0;
}
```

二、改错型实验

（1）功能：编制统计营业员一天的营业额的程序。设程序采用一个循环实现，每次循环输入一笔交易金额并累计营业额。由于营业员一天完成的交易次数是不确定的，为此以最后附加输入一笔0或负数交易额作为交易金额已全部输入结束的标志。以下为统计营业员一天营业额的算法：

统计营业员一天营业额

```
{
    营业额清 0;
    输入第一笔交易额;
    while(交易额>0.0)
    {
        累计营业额;
        输入下一笔交易额;
    }
```

输出营业额；
}
记一笔交易额为变量 sale，营业额为 sigma。
```
#include <stdio.h>
int main()
{
  /**********FOUND**********/
  float sale,sigma
  sigma=0.0;
  printf("Enter sale data.\n");
  /**********FOUND**********/
  scanf("%f",sale);
  while(sale>0.0)
  {
    /**********FOUND**********/
    sigma+==sale;
    printf("Enter next sale data(<=0 to finish).\n");
    scanf("%f",&sale);
  }
  printf("Sigma of sale is %.2f\n",sigma);
  return 0;
}
```

（2）功能：在键盘上输入一个 3 行 3 列矩阵的各个元素的值（值为整数），求输出矩阵第一行与第三行元素之积，并在 fun()函数中输出。

```
#include <stdio.h>
int fun(int a[3][3])
{
  int i,j,sum;
  /**********FOUND**********/
  sum=0;
  /**********FOUND**********/
  for(i=0;i<3;i++)
    for(j=0;j<3;j++)
      /**********FOUND**********/
      sum=*a[i][j];
  return sum;
}
int main()
{
  int i,j,s,a[3][3];;
  for(i=0;i<3;i++)
  {
    for(j=0;j<3;j++)
      scanf("%d",&a[i][j]);
  }
  s=fun(a);
  printf("Sum=%d\n",s);
  return 0;
}
```

（3）功能：有一数组内放 10 个整数，要求找出最小数和它的下标，然后把它和数组中最前面的元素即第一个数对换位置。

```
#include <stdio.h>
```

```
int main( )
{
  int  i,a[10],min,k=0;
  printf("\n please input array 10 elements\n");
  for(i=0;i<10;i++)
    /***********FOUND***********/
    scanf("%d", a[i]);
  for(i=0;i<10;i++)
    printf("%d",a[i]);
  min=a[0];
  /***********FOUND***********/
  for(i=3;i<10;i++)
    /***********FOUND***********/
    if(a[i]>min)
    {
      min=a[i];
      k=i;
    }
  /***********FOUND***********/
  a[k]=a[i];
  a[0]=min;
  printf("\n after eschange:\n");
  for(i=0;i<10;i++)
    printf("%d",a[i]);
  printf("\nk=%d\nmin=%d\n",k,min);
  return 0;
}
```

三、设计型实验

（1）功能：从键盘上输入任意实数 x，求出其所对应的函数值。

$$z=(x-4)\text{的二次幂}（x>4）$$
$$z=x\text{ 的八次幂}(x>-4)$$
$$z=z=4/(x*(x+1))(x>-10)$$
$$z=|x|+20(\text{其他})$$

```
#include <math.h>
#include <stdio.h>
float y(float x)
{
  /**********Program**********/

  /**********  End  **********/
}
int main()
{
  float x;
  scanf("%f",&x);
  printf("y=%f\n",y(x));
  return 0;
}
```

（2）功能：求 1 到 100 之间的偶数之积。

```
#include <stdio.h>
```

```
double  fun(int m)
{
  /**********Program**********/

  /**********  End  **********/
}

int main()
{  printf("ji=%f\n",fun(100));
   return 0;
}
```

提高性实验 3

一、填空型实验

（1）功能：从键盘上输入两个复数的实部与虚部，求出并输出它们的和、差、积、商。

```
#include<stdio.h>
int  main()
{
  float a,b,c,d,e,f;
  printf("输入第一个复数的实部与虚部：");
  scanf("%f, %f",&a,&b);
  printf("输入第二个复数的实部与虚部：");
  scanf("%f, %f",&c,&d);
  /***********SPACE***********/
  【?】;
  f=b+d;
  printf("相加后复数：实部：%f,虚部：%f\n",e,f);
  e=a*c-b*d;
  /***********SPACE***********/
  【?】;
  printf("相乘后复数：实部：%f,虚部：%f\n",e,f);
  e=(a*c+b*d)/(c*c+d*d);
  /***********SPACE***********/
  【?】;
  printf("相除后复数：实部：%f,虚部：%f\n",e,f);
  return 0;
}
```

（2）功能：有 n 个人围成一圈，按顺序排号。从第一个人开始报数（从 1 到 3 报数），凡报到 3 的人退出圈子，问最后留下的是原来第几号的那位。

```
#include <stdio.h>
#define nmax 50
int main()
{
  int i,k,m,n,num[nmax],*p;
  printf("please input the total of numbers:");
  scanf("%d",&n);
```

```
    p=num;
    /**********SPACE**********/
    for(i=0;【?】;i++)
     /**********SPACE**********/
    *(p+i)=【?】;
    i=0;  k=0;  m=0;  while(m<n-1)
    {
      /**********SPACE**********/
      if(【?】!=0) k++;
      if(k==3)
      {
        *(p+i)=0;
        k=0;
        m++;
      }
      i++;
      if(i==n) i=0;
    }
    /**********SPACE**********/
    while(【?】) p++;
    printf("%d is left\n",*p);
    return 0;
}
```

二、改错型实验

（1）功能：以下程序输出前六行杨辉三角形。

```
            1
         1     1
      1     2     1
   1     3     3     1
1     4     6     4     1
         ……
         ……
#include <stdio.h>
int main( )
{
    static int a[6][6];
    int i,j,k;
    /**********FOUND**********/
    for(i=1;i<=6;i++)
    {
      for(k=0;k<10-2*i;k++)
        printf(" ");
      for(j=0;j<=i;j++)
      {
        /**********FOUND**********/
        if(j==0&&j==i)
          a[i][j]=1;
        else
          /**********FOUND**********/
          a[i][j]=a[i-1][j-1]+a[i][j-1];
        printf(" ");
```

```
      printf("%-3d",a[i][j]);
    }
    /**********FOUND**********/
    printf("\t");
  }
  return 0;
}
```

（2）功能：编写函数，求 2!+4!+6!+8!+10+12!+14!的值。

```
#include <stdio.h>
long sum(int n)
{
  /**********FOUND**********/
  int i,j
  long   t,s=0;
  /**********FOUND**********/
  for(i=2;i<=n;i++)
  {
    t=1;
    for(j=1;j<=i;j++)
    t=t*j;
    s=s+t;
  }
  /**********FOUND**********/
  return(t);
}
int main()
{
  printf("this sum=%ld\n",sum(14));
  return 0;
}
```

（3）功能：编写函数 fun，计算下列分段函数的值。

$$f(x)=\begin{cases} x\times 20 & x<0 \text{且} x\neq -3 \\ \sin(x) & 0\leqslant x<10 \text{且} x\neq 2 \text{及} x\neq 3 \\ x\times x+x-1 & \text{其他} \end{cases}$$

```
#include <math.h>
#include <stdio.h>
double fun(double x)
{
  /**********FOUND**********/
  double y
  /**********FOUND**********/
  if (x<0 || x!=-3.0)
    y=x*20;
  else if(x>=0 && x<10.0 && x!=2.0 && x!=3.0)
    y=sin(x);
  else
    y=x*x+x-1;
  /**********FOUND**********/
  return x;
}
int main()
{
  double x,f;
```

```
  printf("input x=");
  scanf("%f",&x);
  f=fun(x);
  printf("x=%f,f(x)=%f\n",x,f);
  return 0;
}
```

三、设计型实验

（1）功能：编写函数 fun(str, i, n)，从字符串 str 中删除第 i 个字符开始的连续 n 个字符（注意，str[0]代表字符串的第一个字符）。

```
#include <stdio.h>
fun(char str[],int i,int n)
{
  /**********Program**********/

  /********** End **********/
}
int main()
{
  char  str[81];
  int   i,n;
  printf("请输入字符串 str 的值:\n");
  scanf("%s",str);
  printf("你输入的字符串 str 是:%s\n",str);
  printf("请输入删除位置 i 和待删字符个数 n 的值:\n");
  scanf("%d%d",&i,&n);
  while (i+n-1>strlen(str))
  {
    printf("删除位置 i 和待删字符个数 n 的值错！请重新输入 i 和 n 的值\n");
    scanf("%d%d",&i,&n);
  }
  fun(str,i,n);
  printf("删除后的字符串 str 是:%s\n",str);
  return 0;
}
```

（2）功能：实现两个整数的交换。

例如：给 a 和 b 分别输入 60 和 65，输出为：a=65 b=60

```
#include<stdio.h>
void fun(int *a,int *b)
{
  /**********Program**********/

  /********** End **********/
}
int main()
{
  int a,b;
  printf("Enter a,b:");
  scanf("%d%d",&a,&b);
  fun(&a,&b);
  printf("a=%d b=%d\n",a,b);
  return 0;
```

}

提高性实验 4

一、填空型实验

（1）功能：一个自然数被 8 除余 1，所得的商被 8 除也余 1，再将第二次的商被 8 除后余 7，最后得到一个商为 a。又知这个自然数被 17 除余 4，所得的商被 17 除余 15，最后得到一个商是 a 的 2 倍。编写程序，求这个自然数。

```
#include <stdio.h>
int main( )
{
  int i,n,a ;
  i=0 ;
  while(1)
  {
    if(i%8==1)
    {
      n=i/8 ;
      if(n%8==1)
      {
        n=n/8 ;
        /***********SPACE***********/
        if(n%8==7) 【?】 ;
      }
    }
    if(i%17==4)
    {
      n=i/17 ;
      if(n%17==15) n=n/17 ;
    }
    if(2*a==n)
    {
      printf("result=%d\n",i) ;
      /***********SPACE***********/
      【?】 ;
    }
    /***********SPACE***********/
    【?】;
    return 0;
  }
}
```

（2）功能：下面函数为二分法查找 key 值。数组中元素已递增排序，若找到 key 则返回对应的下标，否则返回-1。

```
#include <stdio.h>
fun(int a[],int n,int key)
{
  int low,high,mid;
```

```
      low=0;
      high=n-1;
      /**********SPACE**********/
      while(【?】)
      {
        mid=(low+high)/2;
      if(key<a[mid])
          /**********SPACE**********/
          【?】;
      else if(key>a[mid])
      /**********SPACE**********/
       【?】;
      else
        /**********SPACE**********/
        【?】;
      }
      return -1;
}
int main()
{
  int a[10]={1,2,3,4,5,6,7,8,9,10};
  int b,c;
  b=4;
  c=fun(a,10,b);
  if(c==1)
     printf("not found");
  else
     printf("position %d\n",c);
  return 0;
}
```

二、改错型实验

（1）功能：将 m（$1 \leqslant m \leqslant 10$）个字符串连接起来，组成一个新串，放入 pt 所指字符串中。

例如：把 3 个串 "abc"，"CD"，"EF" 串连起来，结果是 "The result is: abcCDEF"。

```
#include <stdio.h>
#include <string.h>
void fun ( char str[][10], int m, char *pt )
{
  /**********FOUND**********/
  int k, q, i
  for ( k = 0; k < m; k++ )
  {
    q = strlen ( str [k] );
    for (i=0; i<q; i++)
      /**********FOUND**********/
      pt[i] = str[k,i] ;
    /**********FOUND**********/
    pt = q ;
    pt[0] = 0 ;
  }
}
int main( )
```

```
{
    int m, h ;
    char s[10][10], p[120] ;
    printf( "\nPlease enter m:" ) ;
    scanf("%d", &m) ; gets(s[0]) ;
    printf( "\nPlease enter %d string:\n", m ) ;
    for ( h = 0; h < m; h++ ) gets( s[h] ) ;
    fun(s, m, p) ;
    printf( "\nThe result is : %s\n", p) ;
    return 0;
}
```

（2）功能：编写一个程序，计算某年某月有几天（注意要区分闰年）。

```
#include<stdio.h>
int main()
{
  int yy,mm,len;
  printf("year,month=");
  scanf("%d%d",&yy,&mm);
  /**********FOUND**********/
  switch(yy)
  {
    case 1:    case 3:    case 5:    case 7:    case 8:    case 10:
    case 12:    len=31;
        /**********FOUND**********/
        break
    case 4:    case 6:    case 9:    case 11:    len=30;    break;
    case 2:
          if (yy%4==0 && yy%100!=0 || yy%400==0)
            len=29;
          else
            len=28;
          break;
    /**********FOUND**********/
    default
          printf("input error!\n");
          break;
  }
  printf("The length of %d %d id %d\n",yy,mm,len);
}
```

（3）功能：编写一个程序，从键盘接收一个字符串，然后按照字符顺序从小到大进行排序，并删除重复的字符。

```
#include <stdio.h>
#include <string.h>
int  main()
{
  char str[100],*p,*q,*r,c;
  printf("输入字符串:");
  gets(str);
  /**********FOUND**********/
  for(p=str;p;p++)
  {
    for(q=r=p;*q;q++)
       if(*r>*q)
```

```
            r=q;
            /**********FOUND**********/
            if(r==p)
            {
                /**********FOUND**********/
                c=r;
                *r=*p;
                *p=c;
            }
        }
        for(p=str;*p;p++)
        {
            for(q=p;*p==*q;q++);
            strcpy(p+1,q);
        }
        printf("结果字符串: %s\n\n",str);
        return 0;
    }
```

三、设计型实验

（1）功能：编写函数 fun，对主程序中用户输入的具有 10 个数据的数组 a 按由大到小排序，并在主程序中输出排序结果。

```
#include <stdio.h>
int fun(int array[], int n)
{
    /**********Program**********/


    /**********  End  **********/
}

int main()
{
    int a[10],i;
    printf("请输入数组 a 中的十个数:\n");
    for (i=0;i<10;i++)
        scanf("%d",&a[i]);
    fun(a,10);
    printf("由大到小的排序结果是:\n");
    for (i=0;i<10;i++)
        printf("%4d",a[i]);
    printf("\n");
    return 0;
}
```

（2）功能：找出一个大于给定整数且紧随这个整数的素数，并作为函数值返回。

```
#include <stdio.h>
int fun(int n)
{
    /**********Program**********/


    /**********  End  **********/
}
```

```
int main()
{
  int  m;
  printf("Enter m: ");
  scanf("%d", &m);
  printf("\nThe result is %d\n", fun(m));
  return 0;
}
```

提高性实验 5

一、填空型实验

（1）功能：输入学生成绩并显示。
```
# include <stdio.h>
struct student
{
  char number[6];
  char name[6];
  int  score[3];
} stu[2];
void output(struct student stu[2]);
int main()
{
  int i, j;
  /***********SPACE***********/
  for(i=0; i<2; 【?】)
  {
    printf("请输入学生%d的成绩：\n", i+1);
    printf("学号: ");
    /***********SPACE***********/
    scanf("%s", 【?】.number);
    printf("姓名: ");
    scanf("%s", stu[i].name);
    for(j=0; j<3; j++)
    {
      printf("成绩 %d.  ", j+1);
      /***********SPACE***********/
      scanf("%d", 【?】.score[j]);
    }
    printf("\n");
  }
  output(stu);
  return 0;
}
void output(struct student stu[2])
{
  int i, j;
  printf("学号  姓名  成绩1  成绩2  成绩3\n");
  for(i=0; i<2; i++)
```

```
    {
    /**********SPACE**********/
     【?】("%-6s%-6s", stu[i].number, stu[i].name);
     for(j=0; j<3; j++)
      printf("%-8d", stu[i].score[j]);
      printf("\n");
    }
}
```

（2）功能：将字母转换成密码，转换规则是将当前字母变成其后的第四个字母，但 W 变成 A，X 变成 B，Y 变成 C，Z 变成 D。小写字母的转换规则同样。

```
#include <stdio.h>
int main()
{
  char c;
  /**********SPACE**********/
  while((c=【?】)!='\n')
  {
   /**********SPACE**********/
   if((c>='a'&&c<='z')||(c>='A'&&c<='Z'))【?】;
   /**********SPACE**********/
   if((c>'Z'【?】c<='Z'+4)||c>'z')  c-=26;
   printf("%c",c);
  }
  return 0;
}
```

二、改错型实验

（1）功能：一个 5 位数，判断它是不是回文数。例如 12321 是回文数，个位（最低位）与万位（最高位）相同，十位与千位相同。

```
#include<stdio.h>
int main( )
{
 /**********FOUND**********/
 long ge,shi,qian;wan,x;
 scanf("%ld",&x);
 /**********FOUND**********/
 wan=x%10000;
 qian=x%10000/1000;
 shi=x%100/10;
 ge=x%10;
 /**********FOUND**********/
 if (ge==wan||shi==qian)
   printf("this number is a huiwen\n");
 else
   printf("this number is not a huiwen\n");
 return 0;
}
```

（2）功能：利用条件运算符的嵌套来完成此题。学习成绩≥90 分的同学用 A 表示，60～89 分之间的用 B 表示，60 分以下的用 C 表示。

```
#include <stdio.h>
```

```
int main()
{
  int score;
  /**********FOUND**********/
  char *grade;
  printf("please input a score\n");
  /**********FOUND**********/
  scanf("%d",score);
  /**********FOUND**********/
  grade=score>=90?'A';(score>=60?'B':'C');
  printf("%d belongs to %c",score,grade);
  return 0;
}
```

(3) 功能：用插入排序法将 n 个字符进行排序（降序）。

插入法排序的思路是：先对数组的头两个元素进行排序，然后根据前两个元素的情况插入第三个元素，再插入第四个元素……

```
#define N 81
#include <stdio.h>
#include <string.h>
void fun(char *aa, int n)
{
  /**********FOUND**********/
  int a ,b;t;
  for( a = 1; a<n; a++)
  {
    t = aa[a]; b = a-1;
    /**********FOUND**********/
    while((b>=0) and (t>aa[b]))
    {
      aa[b+1]=aa[b]; b--; }
    /**********FOUND**********/
    aa[b+1] = t
  }
}
int main()
{
  char a[N];
  printf("\nEnter a string: ");
  gets( a);
  fun(a , strlen(a));
  printf("\nThe string: ");
  puts(a);
  return 0;
}
```

三、设计型实验

（1）功能：从键盘上输入任意实数 x，求出其所对应的函数值。

$z=(x-4)$的二次幂（$x>4$）

$z=x$ 的八次幂（$x>-4$）

$$z=z=4/(x*(x+1))(x>-10)$$
$$z=|x|+20(其他)$$

```
#include <math.h>
#include <stdio.h>
float y(float x)
{
  /**********Program**********/


  /********** End **********/
}
int main()
{
  float x;
  scanf("%f",&x);
  printf("y=%f\n",y(x));
  return 0;
}
```

（2）功能：将主函数中输入的字符串反序存放。

例如：输入字符串"abcdefg"，则应输出"gfedcba"。

```
#include <stdio.h>
#include <string.h>
#define N 81
void fun(char *str,int n)
{
  /**********Program**********/


  /********** End **********/
}
int main()
{
  char s [N];
  int l;
  printf("input a string:");gets(s);
  l=strlen(s);
  fun(s,l);
  printf("The new string is :");
puts(s);
return 0;
}
```

附录 参考答案

配套教材课后习题参考答案

第1章 C语言概述

1. c .obj .exe
2. 顺序结构 选择结构 循环结构
3. 前100个自然数中奇数之和的算法NS图

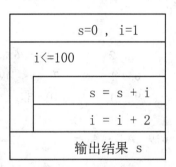

4. 第一行的末尾有多余的分号";"，且stdio.h应该用""或<>括起来
第二行的注释/与*之间不能有空格
Main中的m应该小写
main函数的函数体应该用{}括起来
5. 程序的开始处缺少#include"stdio.h"
    第一行main后缺少（）
    第三行末尾缺少";"，且应把"A"改为"a"
    第四行末尾缺少";"，函数名Printf中的P应改为小写p

第2章 基本C语言程序设计

一、填空题

1. 9	2. 9	3. 30	4. 0	5. 5.6
6. 7	7. 函数	8. 1	9. 9	10. 1.0

11. 双精度   12. 97 b   13. 2.5   14. 循环

二、判断题

1. 对   2. 错   3. 对   4. 错   5. 错
6. 对   7. 对

三、完成程序

1. float x, y, z ;   x, y, z
2. （int）   （int）
3. #define PRICE

四、改错题

1. int h = 5 ; 改为 int h=5, r=5;
   s = 3.14*R*r;改为 s = 3.14*r*r;
2. 加上头文件#include<math.h>
area = sgrt s*(s-a)*(s+b)*(s-c)；改为 area =sqrt(s*(s-a)*(s+b)*(s-c));
3. C1 = "a"； C2 ="b"；改为 c1 = 'a'； c2 ='b'；
Printf ("%C,%C",c1, c2);改为 printf ("%c,%c",c1,c2);
4. int a ; 改为 int a ; long int b ;
printf ("a =%d, b =%f, x =%f, y =%f, u =%u", a, b, x, y, u); 改为
printf ("a =%d, b =%ld, x =%f, y =%f, u =%u", a, b, x, y, u);

第3章　选择结构程序设计

一、填空

1. 325   2. 10   3. 2   4. a=26   b=13   c=19
5. 非零   零   6. ！   7. (x<-4) && (x>4)
8. 3   2   2   9. 1   10. 非   与   或

二、选择

1-5：C D A B B   6-9 A D C A

三、判断

1. 错   2. 错   3. 错

四、完成程序

1. (a>b), (a>c),(b>c)   2. (n % 3 == 0)
3. scanf("%d%d",x,y);   改正：scanf("%d, %d",&x,&y);
if(x<=y)   改正：if (x > y)

五、设计型程序

1.

```
#include<stdio.h>
int main()
{
   float x,y;
printf("Input integer x:");
scanf("%f",&x);
if(x<0 &&x!=-3)    y=x*x+2*x-6;
```

```
       else if(x<10 &&x!=2&&x!=3&&x!=-3)   y=x*x-5*x+6;
            else  y=x*x-x-15;
     printf("y=%f\n",y);
  return 0;
}
```

2.
```
#include<stdio.h>
int main()
{
 int a,b;
 printf("input two number:");
 scanf("%d %d",&a,&b);
 if((a*a+b*b)>=100)  printf("\n %d",(a*a+b*b)/100);
 else            printf("\n %d",a+b);
 return 0;
 }
```

3.
```
# include <stdio.h>
int mth[2][13]={0, 31, 28, 31, 30, 31, 30, 31, 31, 30,31, 30, 31,
     0, 31, 29, 31, 30, 31, 30, 31, 31, 30,31, 30, 31};
isleap(int y)
{  return (y%4 == 0 && y%100 != 0) || (y%400 == 0) ; }
int main ()
{
 int year, month;
 puts("please input year: ");
 scanf("%d", &year);
 puts("please input month: ");
 scanf("%d", &month);
 printf("year %d month %d have %d days。\n", year, month, mth[isleap(year)][month]);
 return 0;
 }
```

4.
```
#include "stdio.h"
int main()
{
   int a;
   printf("Please input data:\n");
   scanf("%d",&a);
   if(a%2)  printf("%d是奇数",a);
   else    printf("%d是偶数",a);
   return 0;
}
```

5. 程序分析：我们想办法把最小的数放到 x 上，先将 x 与 y 进行比较，如果 $x>y$ 则将 x 与 y 的值进行交换，然后再用 x 与 z 进行比较，如果 $x>z$ 则将 x 与 z 的值进行交换，这样能使 x 最小。

```
#include "stdio.h"
int main()
```

```c
{
  int x,y,z,t;
  scanf("%d%d%d",&x,&y,&z);
  if (x>y)  {t=x;x=y;y=t;} /*交换 x,y 的值*/
  if(x>z)   {t=z;z=x;x=t;} /*交换 x,z 的值*/
  if(y>z)   {t=y;y=z;z=t;} /*交换 z,y 的值*/
  printf("small to big: %d %d %d\n",x,y,z);
  return 0;
}
```

6. 程序分析：学会分解出每一位数。

```c
#include <stdio.h>
int main()
{
  long m,a,b,c,d,e;
  printf("请输入一个整数:(0~99999):\n");
  scanf("%ld",&m);
  e=m%10;
  d=m/10%10;
  c=m/100%10;
  b=m/1000%10;
  a=m/10000%10;
  if(m>=10000&&m<=99999)
  {
    printf("你输入的是 5 位数.\n");
printf("逆序输出:%lD. \n",e*10000+d*1000+c*100+b*10+a);
  }
  else if(m>=1000)
  {
    printf("你输入的是 4 位数.\n");
    printf("逆序输出:%lD. \n",e*1000+d*100+c*10+b);
  }
  else if(m>=100)
  {
    printf("你输入的是 3 位数.\n");
    printf("逆序输出:%lD. \n",e*100+d*10+c);
  }
  else if(m>=10)
  {
    printf("你输入的是 2 位数.\n");
    printf("逆序输出:%lD. \n",e*10+d);
  }
  else if(m>=0)
  {
    printf("你输入的是 1 位数.\n");
    printf("逆序输出:%lD. \n",e);
  }
 Return 0;
}
```

7.
```c
#include "stdio.h"
int main()
{
```

```
    int ge,shi,qian,wan,x;
    scanf("%d",&x);
    wan=x/10000;
    qian=x%10000/1000;
    shi=x%100/10;
    ge=x%10;
    if (ge==wan&&shi==qian)   /*个位等于万位并且十位等于千位*/
        printf("this number is a huiwen\n");
    else  printf("this number is not a huiwen\n");
    return 0;
}
```

8. 程序分析：用 switch 语句比较好，如果第一个字母一样，则再用 if 语句判断第二个字母。

```
#include<stdio.h>
int main()
{
    char letter;
    printf("please input the first letter of someday\n");
    while((letter=getchar())!='o')  //输入'o'结束
    {
            //强制键盘缓冲区清空处理,用以清空 while 判断里输入时最后所带的回车
            fflush(stdin);
            switch (letter)
            {
                case 's': printf("please input second letter:\n");
                          if ((letter=getchar())=='a')printf("saturday\n");
                          //直接判断 letter,而不需要再判断 letter=getchar()
                          else if (letter=='u')  printf("sunday\n");
                          else printf("Data error\n");
                          break;
                case 'f':  printf("friday\n");break;
                case 'm':  printf("monday\n");break;
                case 't':  printf("please input second letter:\n");
                          if ((letter=getchar())=='u')printf("tuesday\n");
                          //直接判断 letter,而不需要再判断 letter=getchar()
                          else if (letter=='h')  printf("thursday\n");
                          else  printf("Data error\n");break;
                case 'w':  printf("wednesday\n");break;
                default: printf("error\n");
            }
            fflush(stdin);//强制键盘缓冲区清空处理,清空 case 语句里输入时最后所带的回车
    }
    return 0;
}
```

9. 程序分析：以 3 月 5 日为例，应该先把前两个月的日期数加起来，然后再加上 5 天即本年的第几天，特殊情况，闰年且输入月份大于 3 时需考虑多加一天。

```
#include<stdio.h>
int leapyear(int year)/*是否闰年? */
{
    if(year%400==0)return 1;
        else if(year%4==0&&year%100!=0)return 1;
    return 0;
}
int count(int month,int day)/*不是闰年时是第几天*/
```

```
{
    int a[12]={31,28,31,30,31,30,31,31,30,31,30,31};
    int i,count=0;
    for(i=0;i<month-1;i++)    count=count+a[i];
    return count+day;
}

int main()
{
    int year,month,day,a;
    printf("please input date(yyyymmdd):");
    scanf("%4d%2d%2d",&year,&month,&day);
    a=count(month,day);
    if(leapyear(year)&&month>2)  a=a+1;
    printf("it is the %d day of %d. \n",a,year);
    return 0;
}
```

第4章 循环结构程序设计

一、选择题

1-4：A C D C

二、程序分析题

1. 8 5 2   2. i=3 i=1 i=-1   3. ABABCDCD

三、设计型程序题

1.
```
#include"stdio.h"
int main()
{
  int i,j;
  for(i=1;i<5;i++)
  {
    for(j=1;j<=2*i-1;j++)  printf("*");
    printf("\n");
   }
  for(i=3;i>=1;i--)
  {
    for(j=2*i-1;j>=1;j--)   printf("*");
    printf("\n");
  }
  return 0;
}
```

2.
```
#include"stdio.h"
int main()
{
  int i,j,sum;
  for(i=1;i<=1000;i++)
    {
      for(j=1,sum=0;j<=i/2;j++)
         if(i%j==0)   sum+=j;
      if(sum==i) printf("%-6d",i);
```

```
    }
    return 0;
}
```

3.
```c
#include<stdio.h>
int main()
{
    long f1,f2;
    int i,n,m;
    f1=f2=1;
    printf("输入月数:\n");
    scanf("%d",&n);
    for(i=3;i<=n;i++)
    {
        m=f2;
        f2+=f1;
        f1=m;
    }
    printf("%d\n",f2);
    return 0;
}
```

4.
```c
#include"stdio.h"
int main()
{
    long int num;
    int sum=0;
    int temp;
    printf("please input a number:");
    scanf("%ld",&num);
    while(num>9)
    {
        temp=num%10;
        sum+=temp;
        num/=10;
    }
    sum+=num;
    printf("sum is %d",sum);
    return 0;
}
```

第5章 函数

1. 设计一个函数，求解10000以内的完全数。说明：若一个自然数，恰好与除去它本身以外的一切因数的和相等，这种数叫完全数。例如，6=1+2＋3。

```c
#include"stdio.h"
int p(int n)
{
    int i,sum,flag=1;
    for(i=2,sum=0;i<=n/2;i++)
        if(n%i==0)   sum+=i;
    if(sum!=x)  flag=0;
    return(flag);
```

```
}
int main()
{
    int n,p(int);
    for(n=2;n<=10000;n++)
        if(p(n)==1)   printf("%3d",n);
        return 0;
}
```

2. 设计一个函数 MaxCommonFactor()，利用欧几里德算法（也称辗转相除法）计算两个正整数的最大公约数。

```
#include"stdio.h"
int MaxCommonFactor(int x,int y)
{   int r;
    if (x<y) { r=x;x=y; y=r; }
    do
    { r=x%y;
      x=y;
      y=r;
    }while(r!=0);
    return x;
}
int main( )
{   int m,n, MaxCommonFactor(int,int);
    scanf("%d, %d",&m,&n);
    printf("%d 和%d 的最大公约数为%d",m,n,MaxCommonFactor(m,n));
    return 0;
}
```

3. 利用递归函数求解 Fibonacci 数列问题。Fibonacci 数列表示的是除第 1、2 个数字外，从第三个开始每一个都是前两个之和，例如：1，1，2，3，5，8，13……

```
#include <stdio.h>
#define COL 10
long fibonacci(int n)
{
    if (n==1||n==2)   return 1;
    else  return fibonacci(n-1)+fibonacci(n-2);
}
int main()
{   int i,n;
    long fibonacci(int) ;
    scanf("%d",&n);
    printf("Fibonacci 数列的前%d 项\n", n);
    for (i=1;i<n;i++)
    {   printf("%-10ld",fibonacci(i));
        if(i%COL==0)  printf("\n");
    }
    return 0;
}
```

第 6 章 数组

1. 使用直接插入法完成由小到大排序。

程序分析：直接插入排序(Insertion Sort)的基本思想是：每次将一个待排序的记录，按其关键字大小插入到前面已经排好序的子序列中的适当位置，直到全部记录插入完成为止。

设数组为 a[0…n-1]。
（1）初始时，a[0]自成 1 个有序区，无序区为 a[1…n-1]。令 i=1
（2）将 a[i]并入当前的有序区 a[0…i-1]中形成 a[0…i]的有序区间。
（3）i++并重复第二步直到 i==n-1。排序完成。

程序代码如下：

```c
#include "stdio.h"
int main()
{
  int i,j,temp,num;              /* temp 存储临时数据，num 存储用户输入整数个数 */
  int a[100];                    /*整型数组 a 的长度为 100*/
  printf("Please input the number of integer:\n");
  scanf("%d",&num);
  printf("please input %d  integer number:\n",num);
  for(i=0;i<num;i++)  scanf("%d",&a[i]);   /* 键盘输入 num 个整数 */
  for(i=1; i<num;i++)            /* 直接插入排序 */
  {
      temp=a[i];                 /* 存放当前待插入的元素 */
      j=i-1;                     /* 从前一个位置开始对比，找当前元素合适的位置 */
      while((j>=0)&&temp<a[j])   /* 当前元素比前面的元素小 */
       {
           a[j+1]=a[j];          /* 大的后移 */
           j--;
       }
       a[j+1]=temp;              /* 将当前小的放在前面 */
  }
  printf("sorted result:\n");
  for(i=0;i<num;i++)  printf("%3d ",a[i]);   /* 输出排序后结果 */
  printf("\n");
  return 0;
}
```

2. 随机产生 N 个 0~9 以内的整数，分别统计每个数字出现的次数。

程序分析：本例难点是条件太多，虽然计算机产生数的范围是［0，9］间的 10 个整数，但有 10 种情况。如果用 if 语句或 switch 语句，得判断 10 次，然后对每种不同情况进行计数；再考虑到 10 个变量定义及初值清 0，程序太长太烦。

本例这类问题，我们用数组来解决。设一个只有 10 个元素的数组 time，它的每一个元素 time[i]对应的统计计算机随机产生的数的个数 i，便大功告成。共 N 个随机数，循环次数明确，宜用 for 语句。

程序代码如下：

```c
#include"math.h"
#include"conio.h"
#define N 20
int main()
{
    int time[10],random,i;
    /* 数组 time 存储每个数字出现的个数，random 存放产生的随机数 */
    for(i=0;i<N;i++)  time[i]=0;
    for(i=0;i<N;i++)
    /* 随机产生 N 个 0~9 之间的整数，统计每个数字出现的次数 */
```

```
        {
            random=rand()%10;    /* 产生随机整数存放到变量 random 内 */
            printf("%3d",random); /* 输出变量 random */
            time[random]=time[random]+1; /* 统计每个数字的出现次数,可换成 time[random]++; */
        }
        printf("\n");
         for(i=0;i<10;i++)  printf("\n%d---%d time",i,time[i]); /* 输出结果 */
        return 0;
}
```

3. 数字圈问题。输出可大可小的正方形图案,最外圈是第一层,要求每层上用的数字与层数相同。

例如,当 *N*=5 时,输出:

1 1 1 1 1
1 2 2 2 1
1 2 3 2 1
1 2 2 2 1
1 1 1 1 1

程序代码如下:

```
#include "conio.h"
#include"stdio.h"

#define N 5
int main()
{
    int i,j,k,l,a[N+1][N+1];
    clrscr();
    for(i=1;i<=(N+1)/2;i++)
       for(j=i;j<=N-i+1;j++)
          for(k=i;k<=N-i+1;k++)  a[j][k]=i;
    for(i=1;i<=N;i++)
    {
        for(j=1;j<=N;j++)  printf("%2d",a[i][j]);
        printf("\n");
    }
    return 0;
}
```

4. 由键盘输入一个字符串和一个字符,要求从该字符串中删除指定的字符。

程序分析:

(1)考虑使用两个字符数组 string,temp。其中 string 存放任意输入的一个字符串;temp 存放删除指定字符后的字符串。

(2)设置两个整型变量 i,j 分别作为 string,temp 两个数组的下标,以指示正在处理的位置。

(3)开始处理前 i=j=0,即都指向第一个数组元素。

(4)检查 string 中的当前的字符。

① 如果不是要删除的字符,那么将此字符复制(赋值)到 temp 数组,j 增 1(temp 下次字符复制的位置);

② 如果是要删除的字符,不复制字符,j 也不必增 1(因为这次没有字符复制)。

(5)i 增 1(准备检查 string 的下面一个元素)。

重复第（4）、（5）步直到 string 中的所有字符扫描了一遍。最后 temp 中的内容就是删除了指定字符的字符串。

程序代码如下：

```c
#include"stdio.h"
int main()
{
    char string[20],temp[20],ch;    /* string 存储源串，temp 存储删除某个字符后字符串，
ch 存储被删除字符 */
    int i,j;
    printf("please input string:");
    gets(string);                    /* 键盘输入源串 */
    printf("delete?");
    scanf("%c",&ch);                 /* 键盘输入被删除字符 */
    for(i=0,j=0;i<strlen(string);i++) /* 删除字符的整个过程 */
      if(string[i]!=ch)
      {
          temp[j]=string[i];
          j++;
      }
    temp[j]='\0';                    /* 设置新的字符串结束标志 */
    strcpy(string,temp);             /* 将删除某字符后的字符串复制给 string */
    puts(string);                    /* 输出删除后的字符串 */
    return 0;
}
```

5. 输入一行英文，统计其中有多少个单词（设单词间以空格隔开）。

程序分析：本例难点是单词之间可能有多个空格，单词的个数不等于空格的个数，另外还要考虑到先连续打几个空格后再输入英文的情况。

算法是这样的：设标志变量 flag，在遍历数组元素的过程中，只要遇到空格始终置 flag=0；当遇到非空格并且 flag==0 时（表示它的前一个字符是空格！）证明遇到了一个新单词，单词计数器 k 增 1，同时立即置 flag=1。（如果 flag 不置成 1，当连续遇到非空格字符时，计数器 k 将不断计数而出错。）

程序代码如下：

```c
#include"conio.h"
#include"string.h"
#include"stdio.h"

int main()
{
    char str[80];                    /* 数组 str 存储键盘输入的字符串 */
    int i,count=0,flag=0;/*count 存储单词数，flag 标志*/
    clrscr();
    printf("Please input a string:");
    gets(str);                       /* 键盘输入字符串，存放到 str 数组 */
    for(i=0;(str[i])!='\0';i++)      /* 循环判断字符串中包含的单词数 */
        if(str[i]==' ')  flag=0;     /* 如果当前字符为"空格"，标志置为 0 */
        else
        /* 如果当前字符不是"空格"，并且前一字符是"空格"，则单词计数器加一 */
```

```
        if(flag==0)
          { flag=1; count++; }
    printf("Total words: %d",count);      /* 打印输出字符串中单词个数 */
    return 0;
}
```

6. 随机产生 100 个大小写字母，统计其中各元音字母（不分大小写）的个数。

程序分析：本例难点主要有两点，其一是字母的大小写是随机的；其二是过滤条件太多，需要把复杂问题简单化。

第一个问题的解决可以参考前面章节产生正负数的方法，设一个标志变量 flag，其值 [0,1] 由计算机随机产生，为 0 时对应大写字母，为 1 时对应小写字母。

第二个问题，用 toupper()函数简化条件。

程序代码如下：

```
#include"conio.h"
#include"string.h"
#include"ctype.h"
#include"math.h"
#include"stdio.h"

int main()
{
    char str[100],ch[6]={'\0','A','E','I','O','U'};
    static int flag,i,num[5];     /* 数组 num 存放'A','E','I','O','U'的个数 */
    clrscr();
    for(i=0;i<100;i++)
      {
          flag=rand()%2;                /* 大小写字母标志 */
          str[i]=rand()%26+65;          /* 大写字母 */
          if(flag==0)  str[i]+=32;      /* 小写字母 */
          switch(toupper(str[i]))       /* 统计元音字母个数 */
          {
            case 'A': num[0]++;
            case 'E': num[1]++;
            case 'I': num[2]++;
            case 'O': num[3]++;
            case 'U': num[4]++;
          }
      }
    str[i]='\0';                        /* 赋结束标志 */
    puts(str);                          /* 输出字符串 */
    for(i=1;i<6;i++)                    /* 输出统计结果 */
        printf("\n%c:number:%d\t",ch[i],num[i-1]);
    return 0;
}
```

7. 一维数组 score 内放 10 个学生成绩，定义一个函数求平均成绩。

```
#include"stdio.h"
#define N 10
float average(int a[])
{
    int i;
    float sum=0,ave;
```

```
    for(i=0; i<N; i++)    sum+=a[i];
    ave=sum/N;
    return ave;
}
int main()
{
    float score[N],ave;
    int i;
    printf("Input 10 number:");
    for(i=0; i<N; i++)    scanf("%f",&score[i]);
    ave=average(score);
    printf("The average of 10 students is :%f\n",ave);
    return 0;
}
```

8. 有 M 位学生，学习 N 门课程，已知所有学生的各科成绩，编程实现每位学生的平均成绩和每门课程的平均成绩。

程序分析：根据题意首先定义一个 M 行 N 列的二维数组，存放 M 位学生、N 门课程的成绩，然后按行求和可得每位学生的总成绩，除以课程门数得每位学生平均成绩，按列求和可得每门课程总成绩，除以学生数求得课程平均成绩。该问题可用以下两种方式实现。

程序代码 1：

```
#define Number_student  5
#define Number_course   4
#include "stdio.h"
int main()
{
int i,j;                            /* 循环控制变量 */
static float score[Number_student+1][Number_course+1];
/* 定义二维数组，行数和列数都比学生数、课程数多 1，最后一行用来存放课程平均成绩，最后一列用来存放学生平均成绩 */
clrscr();
for(i=0;i<Number_student;i++)   /* 键盘输入 5 位学生，4 门课程的成绩 */
{
    printf("input student No.%d: 4 course scores:",i+1);
    for(j=0;j<Number_course;j++)    scanf("%f",&score[i][j]);
}
for(i=0;i<Number_student;i++)   /* 计算每位学生的总成绩，每门课程的总成绩、平均成绩 */
{
    for(j=0;j<Number_course;j++)
    {
        score[i][Number_course] += score[i][j];
                /* 求每门课程的总成绩，存放于 score[i][Number_course] */
        score[Number_student][j] += score[i][j];
                /* 求每位学生的总成绩，存放于 score[Number_student][j]*/
    }
    score[i][Number_course]/= Number_course;    /* 求每门课程的平均成绩 */
}
for(j=0;j<Number_course;j++)                    /* 求每位学生的平均成绩 */
        score[Number_student][j] /= Number_student;
printf("\ncomputed result:\n");
printf("student number  course1 course2 course3 course4 average\n"); /* 输出结果 */
for(i=0;i<Number_student;i++)                   /* 输出每位学生的成绩 */
{
```

```c
        printf("  %d\t\t",i+1);                    /* 输出学生编号 */
        for(j=0;j<Number_course+1;j++)             /* 输出每门课程的成绩 */
           printf("%6.1f\t",score[i][j]);
        printf("\n");
    }
    for(j=0;j<8*(Number_course+3);j++) printf("-"); /* 输出一条横线 */
    printf("\ncourse average\t");
    for(j=0;j<Number_course;j++)                   /* 输出课程平均成绩*/
        printf("%6.1f\t",score[Number_student][j]);
    printf("\n");
    getch();
    return 0;
}
```

程序代码 2:

```c
#define Number_student 5
#define Number_course 4
#include "stdio.h"
int main()
{
    int i,j;                                       /* 循环控制变量 */
/* 定义二维数组,行数和列数都比学生数、课程数多 1,最后一行用来存放课程平均成绩,最后一列用来存放学
生平均成绩 */
    static float score[Number_student+1][Number_course+1];
    float average_student,average_course,sum_student,sum_course;
    clrscr();
    for(i=0;i<Number_student;i++)                  /* 键盘输入 5 位学生,4 门课程的成绩 */
    {
      printf("input student %d: 4 course scores:",i);
      for(j=0;j<Number_course;j++)     scanf("%f",&score[i][j]);
    }
    for(i=0;i<Number_student;i++)                  /* 求每位学生的平均成绩 */
    {
      sum_student=0;
      for(j=0;j<Number_course;j++)                 /* 计算第 i 位学生的总成绩 */
         sum_student+=score[i][j];
      average_student=sum_student/Number_course;
      score[i][Number_course]=average_student;
    }
    for(i=0;i<Number_course;i++)                   /* 求每门课程的平均成绩 */
    {
      sum_course=0;
      for(j=0;j<Number_student;j++)                /* 计算第 i 门课程的总成绩 */
         sum_course+=score[j][i];
      average_course=sum_course/Number_student;
      score[Number_student][i]=average_course;
    }
    printf("\ncomputed result:\n");
    printf("student number  course1 course2 course3 course4 average\n"); /* 输出结果 */
    for(i=0;i<Number_student;i++)                  /* 输出每位学生各科成绩、平均成绩 */
    {
      printf("  %d\t\t",i+1);
      for(j=0;j<Number_course+1;j++)    printf("%6.1f\t",score[i][j]);
      printf("\n");
    }
```

```
       for(j=0;j<8*(Number_course+2);j++)  printf("-");      /* 输出一条横线 */
       printf("\ncourse average\t");
       for(j=0;j<Number_course;j++)                /* 输出各门课程的平均成绩 */
            printf("%6.1f\t",score[Number_student][j]);
       printf("\n");
       getch();
       return 0;
}
```

9. 设比赛共有 N 个评委给选手打分，统计时去掉一个最高分和最低分，求选手的最后平均得分。

程序分析：本例难点是找最大数最小数，同时还要排除评委打分全同的情况。本例通过先对选手分数排序，然后去掉首尾两个数，来达到目的。

程序代码如下：

```
#include"conio.h"
#define N 10
int main()
{
   int i,j;
   float score[N],sum=0,average,temp;
/* 数组 score 存储评委打分, sum, average 存储总分和平均分, temp 临时变量 */
   for(i=0;i<N;i++)                /* 键盘录入 N 个评委打分, 并求和 */
   {
       scanf("%f",&score[i]);
       sum+=score[i];
   }
   for(i=0;i<N-1;i++)              /* 将数组 score 各元素按照从小到大排序 */
      for(j=i+1;j<N;j++)
           if(score[j]<score[i])
              { temp=score[i];  score[i]=score[j];  score[j]=temp; }
   average=(sum-score[0]-score[N-1])/(N-2);
    /* 计算去掉一个最高分和最低分的平均分 */
   printf("The end average:%f",average);      /* 输出结果 */
   return 0;
}
```

第7章 指针

1.（答案略）

2.（答案略）

3. 答：数组指针是指向这个数组首地址的指针，指向对象是这个数组；指针数组是存放一类指针的数组，这个数组的每个元素都是一个指针。

4. 解释下面指针说明的含义：

（1）int *p;   p 为指向整型数据的指针变量。

（2）int *p[5];   定义指针数组 p，它由 5 个指向整型数据的指针元素组成。

（3）int (*p)[5];   p 为指向含 5 个元素的一维数组的指针变量。

（4）int *fp();   fp 为返回值是指针的函数，该指针指向整型数据。

（5）int (*fp)();   fp 为指向函数的指针，该函数返回一个整型值。

（6）int * (*fp)(); fp 为指向函数的指针，该函数返回一个指向整型数据的指针。

（7）void *fp();　　fp 为返回值是指针的函数，该指针指向 void 型数据。

（8）int **p;　　　P 是一个指针变量，它指向一个指向整型数据的指针变量。

5. 阅读程序，给出运行结果。

（1）1　2　3　4
　　　5　6　7　8
　　　9　10　11　12

（2）11，11，11

（3）edcba

（4）1　3　5　7

6.（答案略）

7.

```
#include <stdio.h>
int fun(char *s)
{
    int k=0;
    while(*s!='\0')
    {  k++;   s++;  }
    return k;
}
int main()
{
    char str[500];
    printf("String is: \n");
    gets(str);
    printf("长度为:%d",fun(str));
    return 0;
}
```

8.（答案略）

9.

```
#include <stdio.h>
int main()
{
    int day,month,year,sum,leap;
    printf("\nplease input year,month,day\n");
    scanf("%d,%d,%d",&year,&month,&day);
    switch(month)/*先计算某月以前月份的总天数*/
    {
        case 1:sum=0;break;
        case 2:sum=31;break;
        case 3:sum=59;break;
        case 4:sum=90;break;
        case 5:sum=120;break;
        case 6:sum=151;break;
        case 7:sum=181;break;
        case 8:sum=212;break;
        case 9:sum=243;break;
        case 10:sum=273;break;
        case 11:sum=304;break;
        case 12:sum=334;break;
```

```
            default:printf("data error");break;
        }
        sum=sum+day;  /*再加上某天的天数*/
        if(year%400==0||(year%4==0&&year%100!=0))/*判断是不是闰年*/
            leap=1;
        else  leap=0;
        if(leap==1&&month>2)/*如果是闰年且月份大于2,总天数应该加一天*/
            sum++;
        printf("It is the %dth day.",sum);
        return 0;
}
```

10.
```
#include "stdio.h"
char *getMonthName(int month)
{
char*params[]{"January","February","March","April","May","June","July","August",
"September","October","November","December"};
    return params[month-1];
}
int main()
{
    int month;
    char name[10];
    printf("input month(1~12) : ");
    scanf("%d", &month);
    strcpy(name, getMonthName(month));
    printf("%d : %s\n", month, name);
    return 0;
}
```

11.（答案略）

第8章 结构体

1. 正确的算法：
```
#include "stdio.h"
int main()
{
    int z=2;
    struct ABC
{
        char x,xx[10];
        int y,z;
    };
    struct ABC a,b,c;
    scanf("x=%c%c",&a.x);
    gets(b.xx);
    c. z=z+'A';
    printf("%c,%c,%s",c.z,a.x,b.xx);
    return 0;
}
```

2. 分析程序执行结果。

```
num x:     1      2      3,     6
num y:     4      5      6,    15
num z:     7      8      9,    24
Total sum=45
Press any key to continue
```

3. 定义一个结构体 point 表示空间一点。键盘输入空间 N 个点（N 用#define 定义），找出哪一点距原点最远，并输出该点的空间坐标。

```c
#include "math.h"
#include "stdio.h"
#define N 5
typedef struct point{
    //三个坐标成员：x,y,z
    //点到原点举例成员：l
    double x,y,z,l;
}Point;
void InputaddCalcu(Point p[]);
void sortDown(Point p[]);
void output(Point p[]);
int main()
{
    Point points[N];
    InputaddCalcu(points);
    sortDown(points);
    output(points);
    return 0;
}
void InputaddCalcu(Point p[])
{
    //输入N个点的坐标：x,y,z
    int i;
    for(i=0;i<N;i++)
    {
        printf("\n请输入第%d个点的坐标(x,y,z)(空格分隔):",i+1);
        scanf("%lf%lf%lf",&p[i].x,&p[i].y,&p[i].z);
        p[i].l=sqrt(pow(p[i].x,2)+pow(p[i].y,2)+pow(p[i].z,2));
    }
}
void sortDown(Point p[])
{   //按点到原点的距离降序排列数组p
    int i,j;
    Point temp;
    for(i=0;i<N;i++)
        for(j=i+1;j<N;j++)
            if(p[j].l>p[i].l)
            { temp=p[j]; p[j]=p[i];   p[i]=temp; }
}
void output(Point p[])
{//输出点的信息
    int i;
    printf("\n按按点到原点的距离降序输出各点：");
    printf("\n( x , y , z :\t l )");
    for(i=0;i<N;i++)   printf("\n(%.1f,%.1f,%.1f:\t%.1f)",p[i].x,p[i].y,p[i].z,p[i].l);
    printf("\n距离原点最远的点是:");
```

```
    printf("\n(%.1f,%.1f,%.1f:\t%.1f)\n",p[0].x,p[0].y,p[0].z,p[0].l);
    for(i=0;i<N;i++)    //处理最远距离不唯一的情况
        if(p[i].l==p[0].l)printf("\n(%.1f,%.1f,%.1f:\t%.1f)\n",p[i].x,p[i].y,p[i].z,p[i].l);
}
```

4. 定义一个结构体 shangpin 表示一种商品的名称、数量、单价，根据某天商店零售记录。

（1）计算该天总销售额；

（2）按商品名汇总当日销售的商品（某种商品当日销出的总数量）。

```
#include "stdio.h"
#include "string.h"
#define N 50
typedef struct shangpin
{
    char name[20];
    int  num;
    double price;
}ShangPin;
int count=0;
double sum=0.0;
void inputandsum(ShangPin[]);
void sort(ShangPin[]);
void outputInfo(ShangPin[]);
int main()
{
    ShangPin  goods[N];
    inputandsum(goods);
    sort(goods);
    outputInfo(goods);
    return 0;
}
void inputandsum(ShangPin  goods[])
{
    int key;
    printf("\n 请输入商品信息:");
    do{
        printf("\n 商品名:");
        gets(goods[count].name);
        printf("数量:");
        scanf("%d",&goods[count].num);
        printf("单价:");
        scanf("%lf",&goods[count].price);
        sum+=goods[count].num*goods[count].price;
        printf("\n 继续按 0,否则退出!");
        scanf("%d%c",&key);
        count++;
    }while(key==0);
}
void sort(ShangPin  goods[])
{
    int i,j;
    ShangPin temp;
    for(i=0;i<count;i++)
        for(j=i+1;j<count;j++)
            if(strcmp(goods[j].name,goods[i].name)>0)
            {  temp=goods[j];   goods[j]=goods[i];   goods[i]=temp; }
}
```

```
void outputInfo(ShangPin goods[])
{
    int i;
    printf("\n当日商品汇总:");
    printf("\n商品名\t数量\t单价\t总价");
    for(i=0;i<count;i++)
    printf("\n%s\t%d\t%.2f\t%.2f",goods[i].name,goods[i].num,goods[i].price,goods[i].num
*goods[i].price);
    printf("\n总销售额是%.2f\n",sum);
}
```

第9章 文件

一、选择题

1. B 2. C

二、填空题

1. fopen (fname, "w")   ch
2. " r "   (! feof(fp))   fgetc (fp)

三、编程题

1. //以下程序段将输入的N个字符串输出到文件中，在文件中每个字符串占一行。

```
for( i = 1 ; i < N; i++)
{ printf ("Enter a string : "); gets ( s ); fputs( s , fp ); fputc ('\n', fp ); }
//在关闭和重新打开文件后，以下程序段从文件中输入字符串，并将字符串输出到屏幕上。
fgets ( s, M-1, fp );    //从fp所指文件中输入M-1个字符到s的地址中
while (! feof (fp) )
{    c = s[strlen(s) - 1];
    if ( c == '\n') s[strlen(s)-1] = 0 ;
    puts ( s );fgets (s,M-1,fp);
}
```

2. //若数据放到a数组中，以下程序段输出N个双精度数据到已打开的二进制文件中:

```
for( i = 1 ; i < N; i++ ) fwrite (a + i,sizeof (double),1,fp);
//在把文件位置指针移到文件开头之后，可以参考以上程序段从fp所指文件中逐个读入数据//依次存放在a数组中，或用以下语句一次性输入N个sizeof (double)字节的数据，放入a//所指地址开始的存储区。
fread (a,sizeof (double),N,fp);
```

提高性实验参考答案

提高性实验1参考答案

一、填空型实验

（1）答案1：n=strlen(str)或for(n=0;str[n]!='\0';n++)或for(n=0;str[n]; n++)或for(n=0;str[n]!=0;n++);
答案2：i<n 或 n>i 或 n-1>=I；答案3：str[j]>str[j + 1]；答案4：fprintf(fp,"%s\n",str)

（2）答案1：i<=n；答案2：f= -f；答案3：m

二、改错型实验

（1）答案 1：for(i=2;i<=m;i++)；答案 2：y-=1.0/(i*i)；答案 3：return y

（2）答案 1：long ge,shi,qian,wan,x；答案 2：wan=x/10000；

答案 3：if (ge==wan&&shi==qian)

（3）答案 1：int s=1,i；答案 2：if(i%m==0)；答案 3：s*=i

三、设计型实验

（1）
```
int i,j;
    float t,s=0;
    for(i=3;i<=n;i=i+3)
     {t=1;
      for(j=1;j<=i;j++)
        t=t*j;
      s=s+t;}
    return(s);
```

（2）
```
int i,j;
 for(i=0;i<*n;)
   {
    if(bb[i]==y)
       {for(j=i;j<*n;j++)
          bb[j]=bb[j+1];
        *n=*n-1;
        }
       else
         i++;
         }
```

提高性实验 2 参考答案

一、填空型实验

（1）答案 1：str；答案 2：str[i]!='\0'；答案 3：str[k]=str[i]；答案 4：str[k]='\0'

（2）答案 1：float n；答案 2：fabs(t)>=1e-6；答案 3：f=-1*f；答案 4：pi= pi * 4

二、改错型实验

（1）答案 1：float sale,sigma；答案 2：scanf("%f",&sale)；答案 3：sigma+=sale

（2）答案 1：sum=1；答案 2：for(i=0;i<3;i+=2)；答案 3：if (ge==wan&&shi==qian)

（3）答案 1：scanf("%d",&a[i])；答案 2：for(i=1;i<10;i++)；

答案 3：if(a[i]<min)；答案 4：a[k]=a[0]

三、设计型实验

（1）
```
float z;
  if(x>4)  z=sqrt(x-4);
  else if(x>-4)  z=pow(x,8);
  else if(x>-10) z=4/(x*(x+1));
  else z=fabs(x)+20;
  return(z);
```

（2）
```
double   y=1;
    int i;
    for(i=1;i<=m;i++)
      if(i%2==0) y*=i;
      return y;
```

提高性实验3参考答案

一、填空型实验

（1）答案1：e=a+c；答案2：f=a*d+b*c；答案3：f=(b*c-a*d)/(c*c+d*d)

（2）答案1：i<n；答案2：i+1；答案3：*(p+i)；答案4：*p==0

二、改错型实验

（1）答案1：for(i=0;i<6;i++)；答案2：if(j == 0 || j == i);
答案3：a[i][j]=a[i-1][j]+a[i-1][j-1]；答案4：printf("\n")

（2）答案1：int i,j；答案2：for(i=2;i<=n;i=i+2)；答案3：return(s)

（3）答案1：double y；答案2：if (x<0 && x!=-3.0)；答案3：return y

三、设计型实验

（1）
```
while(str[i+n-1])
  {
   str[i-1]=str[i+n-1];
   i++;
  }
  str[i-1]='\0';
```

（2）int t;   t=*a;*a=*b;*b=t;

提高性实验4参考答案

一、填空型实验

（1）答案1：a=n/8；答案2：break；答案3：i++

（2）答案1：low<=high；答案2：high=mid-1；答案3：low= mid + 1；答案4：return   (mid)

二、改错型实验

（1）答案1：int k, q, i；答案2：pt[i] = str[k][i]；答案3：pt += q

（2）答案1：switch(mm)；答案2：break；答案3：default

（3）答案1：for(p=str;*p;p++)；答案2：if(r!=p)；答案3：c=*r

三、设计型实验

（1）
```
int  k, j, t;
  for (k=0;k<n-1;k++)
    for (j=k+1;j<n;j++)
      if (array[k]<array[j])
      {
      t=array[k];
      array[k]=array[j];
      array[j]=t;
      }
```

（2）
```
int i,k;
for(i=n+1;;i++){
for(k=2;k<i;k++)
if(i%k==0)
break;
if(k==i)
return(i);
}
```

提高性实验5参考答案

一、填空型实验

（1）答案1：i++；答案2：&stu[i]；答案3：&stu[i]；答案4：printf

（2）答案1：getchar()；答案2：c+=4；答案3：&&

二、改错型实验

（1）答案1：long ge,shi,qian,wan,x；答案2：wan=x/10000；

答案3：if(ge==wan&&shi==qian)

（2）答案1：char grade；答案2：scanf("%d",&score)；

答案3：grade=score>=90?'A':(score>=60?'B':'C')

（3）答案1：int a, b, t；答案2：while((b>=0)&&(t>aa[b]))；答案3：aa[b+1]=t

三、设计型实验

（1）
```
float z;
  if(x>4) z=sqrt(x-4);
  else if(x>-4) z=pow(x,8);
  else if(x>-10) z=4/(x*(x+1));
  else z=fabs(x)+20;
  return(z);
```

（2）
```
int i,j;
 char c;
for(i=0,j=n-1;i<j;i++,j--)                /*或者 for(i=0,j=n-1;i<n/2;i++,j--)*/
{c=*(str+i);
*(str+i)=*(str+j);
*(str+j)=c;}
```

参考文献

［1］韩忠东. C语言程序设计基础［M］. 北京：电子工业出版社，2007.

［2］谭浩强. C程序设计（第2版）［M］. 北京：清华大学出版社，1999.

［3］严蔚敏. 数据结构（C语言版）［M］. 北京：清华大学出版社，2007.

［4］张静，张继生，白秋颖，唐笑非. C语言程序设计（第2版）上机指导与习题解答［M］. 北京：清华大学出版社，2011.

［5］游洪跃，彭骏，谭斌. C语言程序设计实验与课程设计教程［M］. 北京：清华大学出版社，2011.

［6］鲁云平，周建丽，娄路. C语言程序设计实验教程［M］. 北京：清华大学出版社，2011.

［7］郭有强，周会萍，戚晓明等. C语言程序设计实验指导与课程设计［M］. 北京：清华大学出版社，2009.

［8］教育部考试中心. 全国计算机等级考试二级教程——C语言程序设计. 北京：高等教育出版社，2010.

［9］姜雪，王毅，刘立君. C语言程序设计实验指导［M］. 北京：清华大学出版社，2009.

［10］廖雷. C语言程序设计习题解答与上机指导（第3版）［M］. 北京：高等教育出版社，2010.

［11］耿祥义，李岩. C语言程序设计实训教程［M］. 北京：清华大学出版社，2011.

［12］王萌，唐新来. C语言程序设计实验指导（第1版）［M］. 北京：科学出版社，2009.